Probability Theory and
Computer Science

International Lecture Series in Computer Science

These volumes are based on the lectures given during a series of specially funded chairs. The International Chair in Computer Science was created by IBM Belgium in co-operation with the Belgium National Foundation for Scientific Research. The holders of each chair cover a subject area considered to be of particular relevance to current developments in computer science.

Probability Theory and Computer Science

Edited by

G. Louchard and G. Latouche

Laboratoire d'Informatique Théorique
Université Libre de Bruxelles, Belgium

1983

 **ACADEMIC PRESS**

A Subsidiary of Harcourt Brace Jovanovich, Publishers
London · New York
Paris · San Diego · San Francisco · Sao Paŭlo · Sydney
Tokyo · Toronto

ACADEMIC PRESS INC. (LONDON) LTD.
24–28 Oval Road
London NW1 7DX

U.S. Edition published by
ACADEMIC PRESS INC
111 Fifth Avenue
New York, New York 10003

British Library Cataloguing in Publication Data

Probability theory and computer science.—
  (International lecture series in computer science)
  1. Electronic data processing—Probabilities
  I. Louchard, G.   II. Latouche, G.
  III. Series
  519.2'028'54   QA273

  ISBN 0-12-455820-8

Typeset by Preface Ltd, Salisbury, Wilts.
Printed in Great Britain by Thomson Litho Ltd., East Kilbride, Scotland

Contributors

Donald P. Gaver Naval Postgraduate School, Operations Research Department, Monterey, California 93940, U.S.A.

Hisashi Kobayashi IBM Japan Science Institute, 5–1 Azabudai 3-chome, Monato-ku, Tokyo 106, Japan.

Robert Sedgewick Brown University, Department of Computer Science, Providence, Rhode Island 02912, U.S.A.

Foreword

The present volume gathers together the writings on which Professors Donald P. Gaver, Hisashi Kobayashi and Robert Sedgewick based the series of lectures they presented at the "Université Libre de Bruxelles" in the period from the Fall of 1980 to the Summer of 1981. These lectures took place within the framework of an International Professorship in Computer Science funded by IBM Belgium, and organized by the Belgian National Scientific Research Foundation (FNRS).

The lectures were intended as variations on the central theme "Probability Theory and Computer Science": a theme which had been chosen by Professor Guy Louchard, Head of the Computer Science department of Brussels University, as being in unison with an important part of the research work conducted in this department.

The calculus of probabilities provides a major mathematical tool in the analysis of computer systems and computer programs. But bringing it into play in these types of problems can be a difficult task. The preliminary stages of model building are often tricky owing to the high degree of complexity of the real world of computers and programs. The empirical simplifications of the real problem intricacies, the inference of pertinent analytical characteristics, and the definition of a conceptual model which lends itself to mathematical analysis may require much familiarity with this real world.

Then, when the model has been formulated, the conventional tools and methods of analysis are seldom readily applicable without refinements, generalizations, or more efficient computation techniques to cope with the huge amount of calculations which may be involved. Sometimes such new developments broach fundamental issues in Probability or Stochastic Process Theory; in this respect, the last decade has witnessed the emergence of a closer interaction between computer scientists and applied probabilists.

In the final validation and verification of the analysis, when the practical values of the model must be put to the test for prediction, planning or control purposes, or for a more complete understanding of the real problem under investigation, again much expertise in computers and programming is required.

This classical three-step cyclic process of model inference, analysis and validation is competently and fruitfully laboured in a great variety of applications in each of the three parts of the book.

Besides, very appropriately, each author lays stress on a different step of the process. In the first part, Gaver gives special attention to the construction of stochastic models in these early stages where both imagination and care are required; several problems and techniques are used to show how the basic assumptions of a model can be checked against the observed reality.

Queueing networks and discrete-time queueing processes are fundamental models for the analysis of data traffic in computer and communication systems. Kobayashi, in the second part, surveys and proposes several approaches, exact and approximate, to evaluate efficiently performance measures for these models.

Finally, Sedgewick's material is a very good illustration of how a detailed mathematical analysis can lead to a better understanding of an algorithm, to a simpler and more elegant version of it, and to a more efficient computer implementation.

Each part is self-contained and keeps a fair balance of introductive material. The book is thus accessible to both applied mathematicians and computer scientists, challenging, to paraphrase the last author, the former by its models, and the latter by its mathematical developments. Thanks are due to Professors G. Latouche and G. Louchard who took in charge the practical organization of the lectures and the editing of this book.

Philips Research Laboratory   P. J. Courtois
Brussels, February 1982

Preface

Probability theory and computer science: the former is an ancient science, where names such as Fermat and Pascal play a pre-eminent role—the analysis of games of chance has progressively added to the modern concepts of the theory of probability—the latter led us in thirty years from the 30 tons and 18 000 tubes of ENIAC to the silicon chip and its tens of thousands of transistors per square millimeter.

The connections between the two sciences are ancient and multiple. Pascal himself, in 1642, invented one of the first mechanical adding machines: fifty were built, of which a few may be found in museums, alongside the prototypes of Babbage, and ENIAC itself. These connections formed the theme of the International Professorship in Computer Science, 1980–1981, hosted by the Université Libre de Bruxelles. This professorship has been created by IBM Belgium. The Fonds National de la Recherche Scientifique is entrusted with the organization. Three topics were covered during the courses and seminars: stochastic modeling, queueing system models and the mathematical analysis of combinatorial algorithms.

The first topic is developed by D. P. Gaver in Part I of this book. Gaver emphasizes a modeling attitude, illustrating his subject with numerous examples. Attention is given to problems of, and models for, redundant system reliability and availability, queueing with priorities, first-passage times and areas under path functions of stochastic processes (total waiting time), and various other examples. Also included is a brief account of aspects of modern data analysis, with the implication that its usefulness is significant at the pre-modeling and model-assessment stages of an investigation. Special attention is given in Chapter 3 to distributional sculpturing, an elegant alteration of conventional distributions in order to represent empirical reality more closely.

Queueing system models are examined by H. Kobayashi in Part II. Chapter 4 is devoted to queueing systems that operate on a discrete-time basis: all events are allowed to occur only at regularly spaced points in time. Aside from such discrete structures of an intrinsic nature, it is often computationally convenient to deal with discrete-time systems when one must obtain a numerical solution of a given problem. Some fundamental issues of discrete-time point processes and related queueing systems are treated in detail. In Chapter 5, Kobayashi presents a discussion of diffusion approximations and shows applications to the performance analysis of some simple computer system and communication system models. In Chapter 6 computational algorithms are presented which mark some of the recent progress in the performance analysis of Markovian queueing networks.

Combinatorial algorithms and their mathematical analysis form the subject of Part III by R. Sedgewick. The intersection of fundamental old techniques from mathematics with fundamental new techniques from computer science form an interesting field of study around which a substantial body of knowledge has been built over the past fifteen years. Sedgewick presents algorithms covering a broad range of application areas: algorithms for search in binary trees, permutations, sort, merge, hashing, etc. The mathematical tools, from probability generating functions to asymptotics in the complex plane, are described and applied to these algorithms. The various types of difficulties encountered are clearly identified and illustrated.

It now remains to fulfill the pleasant task of acknowledging the various individuals and institutions whose help and support made possible the organization of this series of lectures. Thanks are due to IBM Belgium for its financial contribution. Mr. t'Kint de Roodenbeke, Directeur des relations extérieures, was particularly helpful in solving organizational problems.

We also wish to thank the FNRS for its permanent effort towards the promotion of research, fundamental and applied, at the highest level. In particular, the effective cooperation of Mr. Levaux, Secrétaire général du FNRS, and Mr. Delers, Secrétaire adjoint, was greatly appreciated.

The three authors have played a crucial role in the success of the International Professorship in Computer Science, 1980–1981. They have shared without restraint their knowledge and expertise with Belgian specialists, from industrial and public organizations as well as from the academic world, giving them a unique opportunity of being brought into contact with new and challenging thoughts.

Brussels   Guy Louchard
June 1982   Guy Latouche

Contents

Part I. Stochastic modeling: Ideas and techniques
Donald P. Gaver

Part II. Stochastic modeling: Queueing models
Hisashi Kobayashi

Part III. Mathematical analysis of combinatorial algorithms
Robert Sedgewick

Contents

PART I

Stochastic Modeling: Ideas and Techniques

Donald P. Gaver

1. Introduction

The primary purpose of these chapters is to summarize the contents of lectures on stochastic modeling presented at the Université Libre de Bruxelles (ULB) in the period March–May, 1981. Much of the material selected for presentation was from the standard menu of probabilistic topics typical of a second course as given to engineers, operations researchers, statisticians, or computer scientists. An attempt was made to emphasize a modeling attitude rather than details of mathematical rigor, illustrating with problems and techniques that are not often prominent in such courses. For example, attention was given to problems of, and models for, redundant system reliability and availability, queueing with priorities, first-passage times and areas under path functions of stochastic processes, (total waiting times), and various other topics. Also included was a brief account of aspects of modern data analysis, with the implication that its usefulness is significant at the pre-modeling and model-assessment stage of an investigation.

A secondary, but gratifying, purpose is to briefly report on cooperative work initiated with the faculty and students at ULB. I wish to mention the enjoyable collaboration with Dr. Guy Latouche, on development of efficient computational methods for repairman-like Markov models in random environments, and with J. P. Colard on the application of sculptured distributions in the simulation evaluation of certain scheduling algorithms. The interest and warm hospitality of Prof. Guy Louchard, Head of the Department of Computer Science at ULB, was also much appreciated.

1.1 THE TOTAL MODELING PROCESS: BRIEF OVERVIEW

It is becoming recognized that the topic of mathematical modeling (including stochastic modeling) exists in its own right as a subject suitable for a formal university course; see Bender (1978). (References for Part I are arranged at the end of Part I.) The modeling step is part of a process of several stages or steps; these may be expressed as follows (Gaver and Thompson, 1973):

(a) identify the general problem area or *situation*; identify specific questions concerning that area;

(b) obtain and analyze subject matter information and data relating to the problem area. Often an examination of such information and data will suggest suitably formulated questions, as in (a);

(c) construct a preliminary model, or models, representing the important features of the situation. Deduce some model implications;

(d) refer the result of (c) to subject matter specialists and decision makers for qualitative critique; revise the model accordingly. This most likely means redoing (a)–(d);

(e) assess the empirical validity of the model to the degree possible. Check the sensitivity of the model conclusions to changes in model assumptions (submodel inputs), and to data variations. Submit to judgement by subject matter experts—but anticipate differences of opinion! The modeling, and remodeling, process may help to reconcile such differences;

(f) compute required answers to interesting questions. Assess the degree of uncertainty in these answers possibly resulting from model misspecification, data bias or other deficiency, computational error, and sampling error in estimates of basic parameters *or* in simulation results used to supply model implications;

(g) communicate, and aid in implementing, the results of the model;

(h) monitor the situation for possible changes in the environment, and hence for the necessity to change the model.

Of course the emphasis in these notes (and in the lectures), was upon the actual modeling step (c). However, some attention was given to the display of data for pre-model examination (Tukey's exploratory data analysis) and to model parameter estimation techniques, particularly those robust methods that attempt to deal with questions of data deficiencies. These latter topics are not, however, included in this chapter.

2. Topics in outline

In this chapter we outline the basic contents of the lectures. These were arranged so as to first present mathematical definitions and properties, and then to illustrate these in terms of sample models for various situations.

2.1 REVIEW OF PROBABILISTIC CONCEPTS, PARTICULARLY CONDITIONING

In this lecture the following basic notions of probability are defined or reviewed: random experiment or trial, sample or event space, events and combinations of events, probability as a function with rules for combination, conditional probability, independence, and Bayes' theorem, random variables and their moments or expectations, transforms (characteristic function, Laplace transform, and generating function) and their moment-generating, convolution, unicity, and continuity properties, plus properties of conditional expectations. In addition, certain classical univariate distributions were reviewed (normal/Gaussian, log-normal, exponential, gamma, etc.)

By way of illustration, a simple problem of equipment (or possibly software) unreliability was considered.

Situation: suppose a system is made up of components that individually fail after a time because of the action of *faults*; the latter may be the result of component misdesign, or attributable to bad installation or adjustment ("human error"), or to a mistake in computer program coding. We wish to relate system failure rate to initial fault content.

Model: N is a random variable (RV) representing the number of faults initially installed unwittingly in the system. Let $\{T_i, i = 1, 2, \ldots, N\}$ be

5

the sequence of RVs describing the failure time of each fault, measured from the time at which the system begins use; T_i may actually be the time at which the service of the particular component is first requested. Suppose the system fails at

$$X_N = \min\{T_1, T_2, \ldots, T_N\} \tag{2.1}$$

Under very simple conditions, namely that all components have the *same* distribution, $F(t)$, of failure time, and all failure times are independent, simple conditional probability arguments yield

$$P\{X_N > t\} = E_N[(1 - F(t))^N] \tag{2.2}$$
$$\equiv g_N(\bar{F}(t))$$

where g_N is the generating function of the number of faults originally sown in the system, and $\bar{F}(t) = 1 - F(t)$ is the survival time distribution per fault. It is easy to see that $P\{X_N = \infty\} = P\{N = 0\}$, possibly >0, so the derived distribution of X_N is quite possibly dishonest. Note that while in general explicit expressions for expectations cannot be obtained (may not even exist), such summaries as the median, 90% point, etc. may be obtained if $g_N(\bar{F}(t))$ can be explicitly inverted, e.g. for N~Poisson and F~exponential. The simplistic assumption of the model may be relaxed, allowing for different T_i distributions, dependence, and so on, and an additional random death time, D applying to the total system can be introduced to induce eventual failure (of physical equipment), or biological death in finite time. There will be less analytical tractability, but simulation may be used to assess system behavior. Statistical estimation problems may be addressed as well; a suitable version of Eq. (2.2) will provide a likelihood function.

Another example of the applicability of a simple conditioning argument is the following.

Situation: when an individual speaks on a telephone or telecommunication channel the conversation is an alternating sequence of talk-spurts and pauses. Similarly, a job being processed on a computer goes through an alternating sequence of CPU (compute) times and IO (input–output) times. Model the total time of the conversation or job processing time, and particularly the joint distribution of busy and idle segments.

Model 1: let $\{X_i, i = 1, 2, \ldots, K\}$ and $\{Y_i, i = 1, 2, \ldots, K\}$ be the durations of talk-spurts and pauses, respectively, and let K be the number of each. The simplest model assumes $\{X_i\}$ and $\{Y_i\}$ to be independently and identically distributed (IID) sequences of RVs, and themselves to be *conditionally* independent, given K, also a RV. The joint distribution of total talking (or processing) time, X, and total pause time (IO time), Y, is thus, by simple conditioning,

$$P\{X \leq x, Y \leq y\} = \sum_{k=1}^{\infty} F_X(x)^{k^*} \cdot F_Y(y)^k P\{K = k\} \qquad (2.3)$$

The joint Laplace transform is ($s_1, s_2, \geq 0$)

$$E[e^{-s_1 X} e^{-s_2 Y}] = \sum_{k=1}^{\infty} [\hat{F}_X(s_1)\hat{F}_Y(s_2)]^k P\{K = k\}$$

$$= g_K[\hat{F}_X(s_1)\hat{F}_Y(s_2)] \qquad (2.4)$$

where g_K is the generating function of K. Put $s_1 = s_2 = s$ to recover the transform of $X + Y = L$, the total conversation length. In the case where $X_i \sim \mathrm{expon}(\lambda)$ and $Y_i \sim \mathrm{expon}(\mu)$ and $K \sim \mathrm{geom}(\alpha)$ are independent:

$$E[e^{-s_1 X} e^{-s_2 Y}] = \sum_{k=1}^{\infty} \left(\frac{\lambda}{\lambda + s_1}\right)^k \left(\frac{\mu}{\mu + s_2}\right)^k (1 - \alpha)\alpha^{k-1}$$

$$= \frac{\lambda\mu(1 - \alpha)}{(\lambda + s_1)(\mu + s_2) - \lambda\mu\alpha} \qquad (2.5)$$

and

$$E[X] = [\lambda(1 - \alpha)]^{-1}, \qquad E[Y] = [\mu(1 - \alpha)]^{-1} \qquad (2.6)$$

Furthermore ($\bar{\alpha} = 1 - \alpha$)

$$E[L] = E[X] + E[Y] = \frac{\lambda + \mu}{\lambda\mu\bar{\alpha}} \qquad (2.7)$$

and

$$\mathrm{Var}[L] = (E[L])^2 - \frac{2}{\lambda\mu\bar{\alpha}} \qquad (2.8)$$

Notice also that the mechanism of *randomization of a sum*, or *mixing* (see Feller, 1966) which has given (2.3) may be used to generate families of bivariate (multivariate) exponential distributions for other modeling purposes.

Model 2: a plausible alternative to the above model assumes X_i and Y_i are not independent, being possibly positively correlated—a long talk-spurt tending to result in a long pause (response by conversationalist). Most simply, $Y_i = \beta X_i$, $\beta > 0$. Then again transform in the $X \sim \mathrm{expon}(\lambda)$ case to get

$$E[e^{-s_1 X} e^{-s_2 Y}] = \frac{\lambda(1 - \alpha)}{s_1 + \beta s_2 + \lambda(1 - \alpha)}$$

or, if $\beta = \lambda/\mu$, which preserves the marginals of Model 1,

$$E[e^{-s_1 X}e^{-s_2 Y}] = \frac{\lambda(1 - \alpha)}{s_1 + \dfrac{\lambda}{\mu}s_2 + \lambda(1 - \alpha)} = \frac{\lambda\mu\bar{\alpha}}{s_1\mu + s_2\lambda + \lambda\mu\bar{\alpha}}. \quad (2.9)$$

It is now immediate that the marginal distribution of $X \sim \text{expon}(\lambda\bar{\alpha})$, $Y \sim \text{expon}(\mu\bar{\alpha})$ and that now the df of the total time, L, is simply $\text{expon}[\lambda\mu\bar{\alpha}/(\lambda + \mu)]$—a much simpler form than that occurring in Model 1 above, which involves a Bessel function. The variance of L in Model 2, being $(E[L])^2$, is also larger than that for Model 1, see (2.8), suggesting that the former model has a longer tail, hence predicting a greater proportion of extremely long conversations.

The above illustrates that the same situation can easily give rise to two—or more—different models, depending upon the manner in which stochastic assumptions are introduced. At best, the introduction should be guided by observed data; at least, sensitivity analyses using different assumptions can outline the range of specification uncertainty.

2.2   MODELS INVOLVING REPEATED TRIALS

A great many situations may be initially modeled in terms of repeated independent trials, where this means that on each of a possibly countably infinite number of occasions a trial (or experiment, or observation) is performed, with outcome X_i (possibly a vector RV) on the ith trial; $\{X_i, i = 1, 2, \ldots\}$ are IID RVs. *Bernoulli trials* are a prime example: flip a biased coin indefinitely; let I_i be one if a head (success) results, and zero if a tail (failure) otherwise, and assume that the probability of success on any trial is independent of all previous outcomes. Equally X_i may be the winnings on a bet at occasion i, with X_i in dollars and either positive or negative. Or X_i may even represent the increase or decrease in a common stock price on the New York Stock exchange, according to some observers.

It is convenient, but somewhat more questionable, to adopt the repeated trial model for modeling real operational and physical phenomena, yet it is often done uncritically. For instance, the lifetimes (times between failure) of computing equipment are often modeled by IID RVs, repair times likewise, queueing system interarrivals and service times as well, inventory demand sizes, sizes of deposits of resources (petroleum) as well, . . . the list is very long, and observational support for these assumptions is usually conspicuously lacking. The attraction of the repeated trials model is mainly its mathematical tractability, which leads to elegant and appealing results.

Often a brief data analysis in terms of marginal distributions of observed X_i seems to provide justification. The simple repeated trials model cannot well represent, say, systematic daily changes in job interarrival times, or numbers of jobs per hour, at a computer center, seasonal effects on such computer system demand, or the influence of other variables such as the introduction of a new class of computer users upon a measure of computer loading. Also, the model does not well describe the sequence of daily rainfalls in a region, nor many other environmental variables. Some examples follow in which repeated trial models seem intially plausible, but which doubtless can stand improvement.

Situation: a structure is to be designed to withstand (wind-generated, seismic, or other) shocks. Question: how long will it survive the environmental stresses, given its initial strength Z?

Model: let the structure have strength Z (suitable units). If the ith year's maximum shock is X_i, assume $\{X_i\}$ to be IID with df $F_X(x)$. Then the time T until the structure failure exceeds t ($t = 1, 2, \ldots$) if $X_i < Z$, $\forall i = 1, 2, \ldots, t$, so that

$$P\{T > t \,|\, Z\} = [F_X(Z)]^t, \qquad (2.10)$$

the geometric distribution, while if Z itself is regarded as random

$$P\{T > t\} = E_Z P\{T > t \,|\, Z\} = E_Z\{[F_X(Z)]^t$$
$$\neq (E_Z\{[F_X(Z)]\})^t; \qquad (2.11)$$

the unconditional distribution is a convex (probability) combination of exponentials. It will resemble an exponential, but often has an extended right tail. It is obviously important that the condition on equipment stress, Z, be removed at the appropriate stage—at the end. Removal of the condition before each "strength test" would be appropriate only if the structure were completely repaired or replaced with another having the unchanged distribution of Z *before* each trial.

Note that assuming yearly maximum environmental events to be IID is intuitively plausible, but some statistical evidence exists for truly long-range correlation in weather data that may call even this assumption into question.

The Bernoulli counting process is an important special case of repeated trials, on each of which either success (probability p) or failure (probability $q = 1 - p$) occurs; $\{N(t), t = 1, 2, \ldots\}$ is the number of successes in $(1, t])$. Times (number of trials) τ between successive successes are IID and geometrically distributed ($P\{\tau = k\} = q^{k-1}p$). The number $N(t)$ of successes in fixed time is binomial (p, t). and is in turn approximately normal (tp, tpq) as $t \to \infty$. Of course Bernoulli trials describe the outcomes of many other repeated trial situations; for instance, the number of jobs submitted to a processing facility requiring more than x time units of processing may be modeled as a Bernoulli counting process with $p = 1 - F_X(x)$, F_X being the distribution of job processing time.

Generalizations of Bernoulli trial situations may be (a) to variations of success probability with trial number ("time", or "space"): $P\{X_i = 1\} = p_i$. Here $E[N(t)] = \sum_{i=1}^{t} p_i$ and $\mathrm{Var}[N(t)] = \sum_{i=1}^{t} p_i q_i \leq t\bar{p}(1 - \bar{p})$, where $\bar{p} = 1/t \sum_{i=1}^{t} p_i$, so variability is *under-represented* if trial-to-trial probabilities change, but a normal approximation to $N(t)$ may still hold. A second generalization is to (b) independently randomize p in the binomial distribution, say according to a beta distribution, creating the beta–binomial distribution. This has been found useful in reliability modeling and in Bayesian inference. A third and important elaboration (c) is to develop a statistical regression model for success probability p based conveniently on the *logistic function*:

$$p_i = \frac{e^{\alpha + \beta u_i}}{1 + e^{\alpha + \beta u_i}} ; \qquad (2.12)$$

here u_i (possibly a vector) represents the influence of other factors upon success probability. Given observations of the form (I_i, u_i), where $I_i = 1$ indicates success on trial i, one can estimate α and β (vector) by maximum likelihood; see Cox (1969). Generalizations to multiple-category situations are possible, and computational methods for parameter estimation and model assessment have been devised; see Pregibon (1981).

The distribution of the number of counts $N(t)$ in t trials of a Bernoulli trials process can be computed by making use of a *forward equation*. Let

$$P_j(t) = P\{N(t) = j \mid N(0) = 0\}$$

Then

$$P_j(t) = P_j(t - 1) \cdot (1 - p) + P_{j-1}(t - 1) \cdot p; \qquad (0 \leq j \leq t) \ (2.13)$$

on the basis of conditioning on events that have happened up to $t - 1$. One can generalize to nonstationary success probabilities easily:

$$P_j(t) = P_j(t - 1) \cdot (1 - p_t) + P_{j-1}(t - 1)p_t \qquad (2.14)$$

Initial conditions may be

$$P_0(0) = 1, \qquad P_j(0) = 0, \qquad j = 1, 2, \ldots$$

A further generalization allows success probability to depend upon the number of previous successes; then

$$P_j(t) = P_j(t - 1)(1 - p_{j,t}) + P_{j-1}(t - 1)p_{j-1,t} \qquad (2.15)$$

and the distribution $P_j(t)$ can easily be computed recursively given the success probabilities

$$p_{j,t} = P\{N(t) = j + 1 \,|\, N(t) = j\} \qquad (2.16)$$

This is a preview of ideas of Markov chains, to be treated later. These expressions are introduced to suggest early on that the answers to interesting and comparatively complex problems can be directly *computed numerically* (in this case iteratively, starting from the initial conditions). Closed-form expressions such as the binomial distribution are handy, and normal approximations are even handier, but one need not modify the facts merely for the sake of convenience.

2.3 SUMS OF REPEATED TRIAL (IID) RANDOM VARIABLES; "LARGE DEVIATIONS"

Models for total demand for physical inventory or for facility (computer) time often naturally involve sums of varying components, modelled as RVs; thus total demand from n sources, or over n time periods, is

$$S_n = X_1 + X_2 + \ldots + X_n \qquad (2.17)$$

If X_i is the (dollar) profit in the ith year for some enterprise, then a financial measure of success is

$$S_n(r) = \sum_{i=1}^{n} X_i r^i \qquad (2.18)$$

where r is a discount rate $(0 < r \le 1)$.

Situation: a computer center experiences varying monthly demands, X_i for the ith month. Here are answers to several simple questions involving sums of X_is.

The expected yearly (n-period) demand is

$$E[S_n] = \sum_{i=1}^{n} E[X_i], \qquad (2.19)$$

the sum of the expected monthly demands, and also

$$\mathrm{Var}[S_n] = \sum_{i=1}^{n} \mathrm{Var}[X_i] \qquad (2.20)$$

provided the X_is are *uncorrelated*. Importantly, as $n \to \infty$

$$F_{S_n'}(x) \equiv P\left\{\frac{S_n - E[S_n]}{\sqrt{\text{Var}[S_n]}} \leq x\right\} \approx \int_{-\infty}^{x} e^{-1/2z^2} \frac{dz}{\sqrt{2\pi}} \equiv \phi(x), \quad (2.21)$$

i.e. S_n becomes approximately normally distributed by the Central Limit Theorem no matter what the distributions of X_i, provided the X_i components are all of about the same size (certainly if they all come independently from the same parent distribution with finite mean and variance). For smallish n and distinctly non-normal components the approximation is improved by an Edgeworth expansion (Feller, 1966), wherein for the equal component example

$$F_{S_n'}(x) = \phi(x) + \left(\frac{\mu_3}{\sigma^3}\right) \cdot \frac{1}{6\sqrt{n}}(x^2 - 1)\frac{d\phi}{dx} + R_n \quad (2.22)$$

where $R_n = 0(1/\sqrt{n})$, and the components are assumed to have densities. Here $\mu_3 = E[(X - E[X])^3]$ and the term $\mu_3/\sigma^3 = \gamma_1$ is the conventional dimensionless *skewness measure* for a distribution, being zero for symmetric distributions (normal), and being $+2$ for the exponential distribution. Additional terms involving kurtosis (4th moments) improve the approximation, but it is possible that Edgeworth numerical values can be "infeasible": the approximation can actually *decrease* with x in certain ranges. Nevertheless the Edgeworth series has been usefully applied, even to unequal component situations, for estimating the loss of capacity of an electric utility; see Levy and Kahn (1981).

A useful alternative is the *method of large deviations*; Feller (1966), Daniels (1954), and others. The ingenious idea is to tilt (or sculpture) the df components so as to make a normal approximation more effective at predicting the probability that $S_n > x$ for large x. For equal components with df $F(x)$ look at the tilted probability measure (assumed to exist for $s > 0$, which sometimes restricts the theoretical applicability):

$$V(dx) = \frac{e^{sx}F(dx)}{e^{\psi(s)}}, \quad (2.23)$$

$\psi(s) = \ln \hat{F}(s) = \ln E[e^{sX}]$ being a cumulant generating function for F, or X. Manipulations show that

$$P\{S_n > z\} \equiv \int_z^{\infty} F^{n*}(dz) = e^{n\psi(s)}\int_z^{\infty} e^{-sx}V^{n*}(dx), \quad (2.24)$$

and the idea is to approximate V^{n*} by a normal distribution centered at z, a feat that can be accomplished by choice of s. It turns out that it is necessary to solve (sometimes numerically) for $s(z)$ the equation

$$z = n\psi'(s) \tag{2.25}$$

in order that the mean of the approximating normal distribution be at z; the variance is $n\psi''(s)$. Finally,

$$P\{S_n > z\} \approx \exp n\{\psi[s(z)] - s(z)\psi'[s(z)] + 1/2s^2(z)\psi''[s(z)]\} \times$$

$$\times \frac{1}{\sqrt{2\pi}} \int_{s(z)\sqrt{n\psi''[s(z)]}}^{\infty} e^{-1/2v^2} \, dv. \tag{2.26}$$

The above technique can also be applied to a compound Poisson model (n is replaced conditionally by $N(t)$, the counting process of a Poisson process, and the condition then removed). Such models are frequently employed in inventory studies; apparently the large deviation approximation has not been applied in that area.

2.4 BERNOULLI TRIALS AND THE POISSON PROCESS: RARE EVENTS

Bernoulli trials (BT) are a special case of the repeated trials model, with events occurring ("success") or not being permitted to occur ("failure") at specific integer time points, often equally spaced. In practice the fixed intervals between trials may be largely arbitrary, and it is attractive to think of events occurring at any (real-valued) time; from this comes the Poisson process (PP). One approach to the PP properties is to consider a BT process to operate over time t with unit time steps, and then refine the time steps (e.g. let $t = 1$ day and starting with possible demands at 15 min intervals, then down to 7.5 min, then to 3.75 etc.) to create a sequence of BT models. The limit of the sequence after ultimate refinement describes the PP.

Specifically, let $T(k)$ be the generic time between successes in the (kth) BT model with time steps $1/2^k$ ($k = 0, 1, 2, \ldots$). This means that the actual number of steps in time t for the BT model k is $2^k t$; correspondingly, let the probability of success per step be $p/2^k$. By conditioning on the first step's outcome this means that

$$E[T(k)] = 1/2^k + (1 - p/2^k) \cdot E[T(k)] \tag{2.27}$$

so $E[T(k)] = 1/p$ for every model, as should be true. Furthermore, as $k \to \infty$ so time steps become arbitrarily small,

$$P\{T(k) > t\} = (1 - p/2^k)^{t \cdot 2^k} \to e^{-tp} \tag{2.28}$$

inter-event times become exponentially distributed and independent in the

(PP) limit. Furthermore the number of PP events ("successes") in time t have the Poisson distribution.

The PP is usefully invoked for many modeling purposes.

Situation: consider a sequence of days on which demands for computer service (time) are made, and focus on the occurrence patterns of *runs* (uninterrupted sequences) of *high-demand* days. Question: what is the distribution of times between successive runs, and what is the distribution of the number of such runs in a fixed time t? It will turn out that, if either the run lengths are long, or if the probability of a high-demand day is small, that runs tend to occur as a Poisson process if the time scale is appropriate.

Model: begin by modeling individual high demand day occurrences as successes in BT. Let $\tau_1(k)$ represent the time until the 1st occurrence of a run of length k, and, measured from the end of such a run, let $\tau_2(k)$, $\tau_3(k), \ldots \tau_i(k), \ldots$ be the time until the 2nd, 3rd, $\ldots$, ith, such run is realized. By the BT assumption $\{\tau_i(k), i = 1, 2, \ldots\}$ is an IID sequence of RVs. Then we can represent $\tau(k)$ by conditioning on the events that may occur in the first k trials:

$$\tau(k) = \begin{cases} k & \text{with probability} \quad p^k \\ 1 + \tau'(k) & \text{with probability} \quad q \\ \cdots \\ j + \tau'(k) & \text{with probability} \quad p^{j-1}q \; . \\ \cdots \\ k + \tau'(k) & \text{with probability} \quad p^{k-1}q \end{cases} \qquad (2.29)$$

where $\tau'(k)$ is an independent replica of any $\tau(k)$: the idea is that the process *starts over* once a failure occurs to spoil a run. Alternatively,

$$\tau(k) = \begin{cases} k & \text{with probability} \quad p^k \\ R(k) + \tau'(k) & \text{with probability} \quad 1 - p^k \end{cases} \qquad (2.30)$$

where

$$P\{R(k) = j\} = \frac{qp^{j-1}}{1 - p^k}, \qquad j = 1, 2, \ldots, k, \qquad (2.31)$$

a truncated geometric distribution. From these come the generating function of $\tau(k)$, and in principle its distribution:

$$E[z^{\tau(k)}] = \frac{z^k p^k}{1 - qz\left[\dfrac{1 - (pz)^k}{1 - pz}\right]} \qquad (2.32)$$

Differentiation gives the mean

$$E[\tau(k)] = k + \frac{1}{p^k}\left[\frac{1}{1-p} - \frac{kp^k}{1-p^k}\right] \approx \frac{1}{p^k(1-p)} \qquad (2.33)$$

the approximation holding if either $p \to 0$ or $k \to \infty$. In either case the run is a rare event.

While explicit inversion of the expression for $E[z^{\tau(k)}]$ is possible by use of partial fractions, the result is quite complicated. On the other hand, look for the distribution of

$$\tau^*(k) = \tau(k)/E[\tau(k)] \qquad (2.34)$$

when $E[\tau(k)]$ becomes large. The expectation

$$E[e^{-s\tau^*(k)}] = E[e^{-s\tau(k)/E[\tau(k)]}] \qquad (2.35)$$

can be obtained from the generating function (by putting $z = \exp[-sp^k q]$); next let either $p \to 0$ (rare individual events) or $k \to \infty$ (long runs) to find that this transform converges to $(1 + s)^{-1}$. Then by the unicity theorem for transforms (Feller, 1966) the normalized RV $\tau^*(k)$ is approximately unit exponentially distributed, i.e.

$$P\{\tau(k) \le tE[\tau(k)]\} \approx 1 - e^{-t} \qquad (2.36)$$

and furthermore the distribution of the numbe of k-runs in time t, $N_k(t)$, is approximately Poisson. Deviation from the Poisson (indicated by over-variance) may signify that the underlying demand generating process is inhomogeneous or cluster-prone in time, and that extra facilities are required to reduce backlogs. Examination of runs is one way to check the validity of the basic modeling assumption of Bernoulli trials.

Similar limiting arguments simplify other situations involving rare events that are generated by even more complicated processes. (See work on first-passage times for combinations of random loads by Gaver and Jacobs (1981).

2.5 MARKOV MODELS: GENERAL COMMENTS

The basic theory of Markov chains and processes, both in discrete and continuous time, is well introduced in standard texts such as Feller (1966), Chung (1967), Karlin and Taylor (1975) Kleinrock (1976), and needs no systematic coverage, only review and illustration. By way of review, recollect the ideas of various possible state space definitions: integers, integer and real numbers ("ages"), real numbers (e.g. virtual waiting times in queues); times (index sets) either discrete and equally-spaced or imbedded or continuous time; Markov property defined by conditional probabilities ("The future is independent of the past, given the present"). Carry on to matrix representation of the state probabilities after t (0, 1, 2, ...) time steps, forward and backward Chapman–Kolmogorov equations, generalize to a discrete state Markov chain in continuous time with exponential sojourns in states, state classification emphasizing irreducible chains and transient chains (with at least one absorbing barrier), recurrent events and first-passage times and absorption probabilities, generating functions and other transforms.

Simple Markovian assumptions i.e. that a scalar state RV $X(t)$, where t is time or space, is Markov, introduce dependence in a plausible and tractable manner. Usually it is necessary to assume, for example, that the one-step transition probabilities (discrete state, discrete time)

$$p_{ij} = P\{X(t) = j \mid X(t-1) = i\} \tag{2.37}$$

are time-homogeneous in order to obtain explicit neat solutions. Analogous assumptions must be made about discrete-state Markov processes in continuous time, wherein λ_i is the rate of departure from state i (exponentially distributed sojourn time parameter), and p_{ij} is the corresponding probability of a move from i to destination j. Of course a known deterministic time dependence, involving daily or weekly cycles, and trends can be dealt with by numerically multiplying the transition probability matrices.

More irregular changes in process behavior can be represented as the effect of randomly changing external events, or *random environments* for short. In such models the actual primary process transition parameters (e.g. p_{ij}, or λ_i) change in time under the influence of such environmental factors as seismic vibration, temperature and humidity, ocean sea state, wind speed or other meteorological effects, or variations in personnel effectiveness and propensity for errors. Random environment models conveniently postulate that environmental changes induce simple discrete-state Markovian behavior on the basic or primary process parameters; of most interest are parameter changes that occur considerably more slowly than do state changes in the basic process.

Markov modeling of real situations usually involves simplifications at certain crucial stages. Even then, the answers to interesting questions may require extensive computing or simulation. Astute choices of submodels or component models, e.g. the use of "phase-type" distributions for representing arrival and departure processes in queues, can be of help, as can the recognition (or plausible imposition) and exploitation of special structure; see Neuts (1981).

2.6 SOME MARKOV PROCESS PROBLEMS AND MODELS

Here are some illustrative situations and corresponding Markov chain models.

Situation (queueing in discrete time): a servicing facility, e.g. a computer system or a programming (or other) consultant, or a communication channel, experiences single customer arrivals in a random fashion; arrivals enter at the discrete times 0, 1, 2, 3, . . . only, and service completions occur only at such times. Discuss the nature of the delays and backlogs that occur.

Model 1: let the probability of a single arrival at time (epoch) t be $a_i(t)$, where i refers to the number present at that epoch. Each arrival must wait at least one time period before discharge, even if it immediately enters service upon arrival. Let $d_i(t)$ be the probability that an arrival that has been in service at t actually departs at $t + 1$. Now let $X(t)$ denote the number of arrivals in the system which have not yet completed service at time t. Model $\{X(t)\}$ as a Markov chain with the following one-step transition probabilities ($i \geq 1$):

$$p_{i,i+1}(t) = P\{X(t + 1) = i + 1 \,|\, X(t) = i\} = [1 - d_i(t)]a_i(t)$$
$$p_{i,i-1}(t) = P\{X(t + 1) = i - 1 \,|\, X(t) = i\} = d_i(t)[1 - a_i(t)]$$
$$p_{0,1}(t) = a_0(t) \tag{2.38}$$
$$p_{0,0}(t) = 1 - a_0(t)$$
$$p_{ii}(t) = 1 - \{[1 - d_i(t)]a_i(t) + d_i(t)[1 - a_i(t)]\}$$
$$p_{ij}(t) = 0 \text{ otherwise.}$$

If the number in the system is $\leq I$, so the state space is finite,

$$p_{I,I+1}(t) = 0, \qquad p_{I,I-1}(t) = d_I(t)$$

and the probability distribution of $X(t)$ for any t can be obtained by *numerically* multiplying the one-step transition matrices, $\mathbf{P}(t)$, with elements given above:

$$P_{ij}(t) = P\{X(t) = j \,|\, X(0) = i\} = \text{element in } i\text{th row, } j\text{th column of } \mathbf{P}(t)$$

$$= \prod_{t'=0}^{t} \mathbf{p}(t') \tag{2.39}$$

This can be done especially easily in APL if the process is time-homogeneous, i.e. $a_i(t) = a_i$, $d_i(t) = d_i$ independent of elapsed time. Explicit analytical solutions can rarely be found for time-homogeneous cases, let alone for non-time-homogeneous cases. If they were available, the solutions would generally be very complicated and difficult to interpret.

Model 1(a): specialize the above to let $a_i(t) = a > 0$ and $d_i(t) = d > 0$. If the maximum number in the system is I, there is a stationary solution; put $s = a\bar{d}\bar{a}d$, $\bar{a} = 1 - a$, $\bar{d} = 1 - d$:

$$\pi_0 = \frac{(d - a)\bar{d}}{d\bar{d} - a\bar{a}s^I}$$
$$\cdots$$
$$\pi_j = \frac{(d - a)s^j}{d\bar{d} - a\bar{a}s^I} \qquad\qquad (2.40)$$
$$\cdots$$
$$\pi_I = \frac{(d - a)\bar{a}s^I}{d\bar{d} - a\bar{a}s^I}$$

Furthermore, if $s < 1$ then the process tends to drift towards zero and even if there is no upper bound on system states the process is irreducible and ergodic so the long-run distribution is

$$\pi_j = \frac{d - a}{d\bar{d}}s^j, \qquad j = 1, 2, \ldots \qquad\qquad (2.41)$$

a modified geometric distribution, some form of which so often appears in queueing problems. The above simple special case is well known, but can be useful for checking the accuracy of computer programs used to compute numerical solutions to the time-dependent case.

Model 2: let $a_{i,b}(t)$ be the probability that at time t there occurs a *bunch* of arrivals of size b ($b = 0, 1, 2, \ldots$), given that i are awaiting service and will not make further demands. For example, suppose there are I total customers, e.g. computer terminals accessing a central facility, and that each applies for service independently with probability $\alpha(t)$, provided that it is not undergoing service. Then

$$a_{i,b}(t) = \binom{I-i}{b}[\alpha(t)]^b[1 - \alpha(t)]^{I-i-b},$$

and for $k \geqslant 0$,

$$p_{i,i+k}(t) = \bar{d}_i(t)a_{i,k}(t) + d_i(t)a_{i,k+1}(t), \qquad\qquad (2.42)$$

while

$$p_{i,i-1}(t) = a_{i,0}(t)d_i(t)$$

Again the probability distribution of $X(t)$ can be numerically computed.

Model 2(a): suppose the arrivals are caused by a common event (a "common cause" in engineering parlance). This might be the occurrence of an earthquake of large magnitude, or other environmental shock. Let such an event occur at time t with probability $c(t)$; let the probability that b arrivals (demands for service) occur as a result be conditionally binomially distributed with parameter θ. Then

$$a_{i,b}(t) = c(t) \cdot \binom{I-i}{b}\theta^b[1 - \theta]^{I-i-b} \tag{2.43}$$

This can again be used to form one-step transition probabilities, and to calculate state probabilities at any time. The present model allows for a catastrophic shock situation: if $\theta = 1$ then *all* outstanding customers simultaneously demand service, i.e. $I - i$ arrivals occur simultaneously. This differs from Model 2.

Situation (queueing with breakdown of service or preemptive priorities): suppose a single server, e.g. computer facility, or data transmission channel, is confronted by a random arrival stream of basic service demands. These demands may be characterized by their service times, or work request durations such as the times required to transmit single bodies of data or digitized messages. In addition, these services may be effectively prolonged by the occurrence of interruptions, e.g. from internal server breakdowns resulting in temporary processor unavailability, or from environmental noise or even intentional jamming. How is the queue size and waiting time of demands affected by such interruptions? What steps can be taken to reduce the interruption effects?

Model: nearly all classical queueing theory is most conveniently developed if the service times of the individual demands are IID; however, see Jacobs (1978; 1980) for a discussion of a model involving correlation effects. If service times are to be interrupted and *repeated*, or alternatively *resumed*, an interruption process that preserves the IID character of the basic service times allows nearly direct adaptation of conventional theory; such a process is one that requires exponentially distributed ("memoryless") periods between successive interruptions, and this will be assumed. Checks of the sensitivity of results to this reasonable assumption can be made by simulation.

If interruptions of IID duration X (df $F_X(x)$) occur at IID expon (λ_H) intervals, then the time to complete the ith basic service (low priority) is, provided service can *resume* after each service,

$$C_i = S_i + X_1 + X_2 + \ldots + X_{I(S_i)} \tag{2.44}$$

where S_i is the ith basic or low priority service time (df $F_s(x)$), X_j is the duration of the jth interruption, and $I(S_i)$ is the number of interruptions that occur during S_i. Given S_i, $I(S_i)$ is Poisson $(\lambda_H S_i)$, and the Laplace–Stieltjes transform of C_i is, finally,

$$\hat{F}_C(s) = \hat{F}_S[s + \lambda_H\{1 - \hat{F}_X(s)\}] \tag{2.45}$$

and hence

$$E[C] = E[S]\{1 + \lambda_H E[X]\}$$
$$E[C^2] = E[S^2]\{1 + \lambda_H E[X]\}^2 + E[S]\lambda_H E[X^2]. \qquad (2.46)$$

On the other hand, if basic services that are interrupted must begin again from scratch, i.e. services must *repeat*, then the ith completion time becomes

$$C_i = S_i + X_1 + S'_1 + X_2 + S'_2 + \ldots + X_{I(S_i)} + S'_{I(S_i)} \qquad (2.47)$$

where s'_j is the jth interrupted basic service time that must be repeated. The Laplace–Stieltjes transform is

$$\hat{F}_C(s) = E\left\{ \frac{e^{-(\lambda_H + s)S}}{1 - E[e^{-sX}] \dfrac{\lambda_H}{\lambda_H + s}[1 - e^{-(s+\lambda_H)S}]} \right\} \qquad (2.48)$$

and, by differentiation of the latter expression,

$$E[C] = \{E[e^{\lambda_H S}] - 1\}\left\{ E[X] + \frac{1}{\lambda_H} \right\} \qquad (2.49)$$

$$E[C^2] = 2E((e^{\lambda_H S} - 1)^2)\left(E[X] + \frac{1}{\lambda_H} \right)^2 + 2E(Se^{\lambda_H S})\left(E[X] + \frac{1}{\lambda_H} \right)$$

$$+ \left(E(e^{\lambda_H S} - 1)E[X^2] + 2E[X] \cdot \frac{1}{\lambda_H} + \frac{2}{\lambda_H^2} \right)$$

In order to assess queueing delay, look at the process $\{N_d, d = 0, 1, 2, \ldots\}$ describing the number of basic demands at the server at the instants just following departures; $\{N_d, d = 1, 2, \ldots\}$ is an embedded Markov chain provided basic arrivals are (compound) Poisson. Then

$$N_{d+1} = H_d + A(C_{d+1}) \qquad \text{if} \quad N_d = 0$$
$$= N_d + A(C_{d+1}) - 1 \quad \text{if} \quad N_d \geq 1. \qquad (2.50)$$

Here H_d is the number of basic (low-priority) demands made at the beginning of a basic service busy period initiated by the appearances of a high-priority demand, and $A(C_d)$ is the number of basic demands made during the dth basic completion time. Express an arbitrary H as follows

$$H = \begin{cases} 0 & \text{with probability} \quad \dfrac{\lambda_L}{\lambda_L + \lambda_H} \\[2ex] A(C) & \text{with probability} \quad \dfrac{\lambda_H}{\lambda_L + \lambda_H}. \end{cases} \qquad (2.51)$$

It follows by conditional expectations that the embedded chain is ergodic if $E[A(C)] < 1$, and that then the long-run probability of system emptiness at an embedded time point is

$$p_0 \equiv \lim_{d \to \infty} P\{N_d = 0\} = \frac{1 - E[A(C)]}{E[H]} \qquad (2.52)$$

and the long-run expected occupancy is

$$E[N] = \frac{1}{2(1 - E[A(C)])} \times \{E[(H + A(C))^2]p_0 + E[(A(C) - 1)^2](1 - p_0)\} \qquad (2.53)$$

Delay can then be estimated by use of Little's formula. A version of the above formulas correct in continuous time may be found using the results of Gaver (1962); the difference between embedded and continuous time becomes comparatively negligible if the basic traffic intensity $E[A(C)]$ is close to, but below, unity ("heavy traffic"). Since computer system monitoring devices sample the system state at the moment an event occurs (e.g. at a departure instant) a theoretical account of the queue at such moments (imbedded times) is sometimes of direct interest.

An alternative approach to the long-run distribution of delay of an arriving basic demand is by way of Wald's identity or martingales; see Feller (1966). This will actually handle waiting-times when the basic service interarrival intervals are IID, but otherwise arbitrarily, distributed. Still another approach is via the Takacs-type integro–differential equation; see Kleinrock (1976) for an account.

The previous situations have been discussed in terms of long-run probabilities. Frequently questions involving the *time* until system failure (or restitution to operational conditions) are more important.

Situation (redundant repairable systems): a particular system function can be performed if at least k out of n components ("machines") function. For example, electric power is available if at least one generator is working out of two that are installed. Suppose that the system components are all operative initially but fail randomly; failures are immediately detected, but repairs are of random duration, so several machines can be down, all awaiting repair completion. Question: how long will it be until $l = n - k + 1$ components are simultaneously in a failed condition? The time until this occurrence is the time to failure of a k-out-of-n system.

Model 1: it is now convenient (but possibly unrealistic!) to assume that the machines fail independently and after exponentially distributed times in operation, each with rate λ. This simplification may be relaxed, but at the price of expanding the state space. Assume too that the repairs occur in a Markovian manner, e.g. (but not necessarily) at rate $\mu \cdot \min(N(t), R(t))$, where $N(t)$ is the number of machines failed and down for repair at t, and $R(t)$ is the number of repairmen on duty. This is the classical machine repairman problem; see Feller (1966), and Cox and Smith (1961). Usually $R(t) = r$, a constant, although provisions may be made for automatically increasing repair effort when redundancy reserves become dangerously low. In other words, $N(t)$ is a simple birth-and-death Markov process, wherein jumps in state, $N(t)$, occur at exponentially distributed intervals or *sojourn times*, S_i for the sojourn in state i, transitioning always to neighboring states. In the present situation as $\Delta \to 0$

$$P\{N(t + \Delta) = i + 1 | N(t) = i\} = \lambda(n - i)\Delta + o(\Delta)$$
$$= \lambda_i\Delta + o(\Delta)$$
$$P\{N(t + \Delta) = i - 1 | N(t) = i\} = \mu \min(i, r)\Delta + o(\Delta)$$
$$= \mu_i\Delta + o(\Delta),$$

(2.54)

abbreviating the general transition rates to λ_i and μ_i.

In order for system state to reach $i + 1$ from i for the first time it must either do so on the first transition out of state i, or else drop back to $i - 1$, return to i and try again. Thus if U_i is the local first passage time from i to $i + 1$:

$$U_i = \inf\{t: N(t) = i + 1 | N(t) = i\}; \tag{2.55}$$

write

$$U_i = S_i + \begin{cases} 0 & \text{with probability } p_{i,i+1} = \dfrac{\lambda_i}{\lambda_i + \mu_i} \\ U'_{i-1} + U'_i & \text{with probability } p_{i,i-1} = \dfrac{\mu_i}{\lambda_i + \mu_i} \end{cases} \tag{2.56}$$

where U'_i has the same distribution as U_i. The above representation allows immediate derivation of the Laplace–Stieltjes transform of U_i by conditional expectations. The result is

$$E[e^{-sU_i}] \equiv \psi_i(s) = \frac{\lambda_i}{s + \lambda_i + \mu_i[1 - \psi_{i-1}(s)]}, \qquad i = 1, 2, \ldots,$$
$$\psi_0(s) = \frac{\lambda_0}{\lambda_0 + s}$$

(2.57)

Furthermore, since the first-passage time from $N(t) = i$ to $j > i$ can, on the basis of Markovian assumptions, be expressed as

$$T_{ij} = U_i + U_{i+1} + \ldots + U_{j-1} \tag{2.58}$$

where $\{U_{i+k}, k = 0, 1, \ldots\}$ are independent RVs, the Laplace–Stieltjes transform of T_{ij} is

$$E(e^{-sT_{ij}}) = \prod_{k=i}^{j-1} \psi_k(s) \tag{2.59}$$

and the cumulants (moment-like quantities; see Cramér, 1946) can be expressed in terms of those of the U_i. Here are a few moments of the U_i, recursively expressed and hence easily computed:

$$E[U_i] = \frac{1}{\lambda_i}\{1 + \mu_i E[U_{i-1}]\},$$

$$E[u_i^2] = \frac{2}{\lambda_i^2}\{1 + \mu_i E[u_{i-1}]\}^2 + \frac{\mu_i}{\lambda_i}E[U_{i-1}^2],$$

$$E[U_i^3] = \frac{6}{\lambda_i^3}\{1 + \mu_i E[U_{i-1}]\}^3 + \frac{6\mu_i}{\lambda_i^2}\{1 + \mu_i E[U_{i-1}]\}E[U_{i-1}^2]$$

$$+ \frac{\mu_i}{\lambda_i}E[U_i^3], \tag{2.60}$$

and

$$E[U_i^4] = \frac{24}{\lambda_i^4}\{1 + \mu_i E[U_{i-1}]\}^4 + \frac{36\mu_i}{\lambda_i^3}\{1 + \mu_i E[U_{i-1}]\}^2 E[U_{i-1}^2]$$

$$+ \frac{6\mu_i^2}{\lambda_i^2}(E[U_{i-1}^2])^2 + \frac{8\mu_i}{\lambda_i^2}\{1 + \mu_i E[U_{i-1}]\}E[U_{i-1}^3]$$

$$+ \frac{\mu_i}{\lambda_i}E[U_{i-1}^4].$$

From these, standard variance, skewness, and kurtosis measures can be easily computed. For the repairman model discussed initially it can be shown that if the expected time to system failure, $E[T_{0l}]$, is "long" then $T_{0l}/E[T_{0l}]$ resembles an expon(1) RV.

Model 2 (catastrophic failures): suppose that in addition to the independent random failures there is a catastrophic event that "kills" all operative machines simultaneously; let it occur after a time $C \sim \text{expon}(v)$. Then the system failure time T_{0l}^* has a distribution given by

$$P\{T_{0l}^* > t\} = P\{T_{0l} > t\}e^{-vt} \tag{2.61}$$

and from this

$$E[e^{-sT_{0l}^*}] = \frac{1 - E[e^{-(s+v)T_{0l}}]}{s + v} \tag{2.62}$$

from which moments can be generated; see Chu and Gaver (1977). It is sobering to note that if v^{-1}, the mean time to catastrophe occurrence, becomes small or even comparable to $E[T_{0i}]$, then the mean time to redundant system failure is essentially v^{-1}, and redundancy alone may not improve system reliability.

Model 3 (simultaneous repair): if the system is not under constant surveillance, but instead is inspected at random times (rate μ) and then repaired in negligible time, the number of down machines at t may jump essentially instantaneously, either to zero (perfect and rapid repair), or to some lower point (imperfect repair). In this case the basic "skip-free up" character is retained, but now

$$
U_i = S_i + \begin{cases} 0 & \text{with probability } p_{i,i+1} \\ U_i' + U_{i-1}' + U_{i-2}' + \ldots + U_{i-j}' & \text{with probability } p_{i,i-j} . \\ \quad j = 0, 1, 2, \ldots, i & \end{cases} \tag{2.63}
$$

Note that if $j = 0$ then repair is completely ineffective. Conditional expectations now give the Laplace–Stieltjes transform

$$
E[e^{-sU_i}] \equiv \psi_i(s) = \frac{\eta_i(s) p_{i,i+1}}{1 - \eta_i(s) \sum_{j=1}^{i} \left(\prod_{r=1}^{j} \psi_{i-r} \right) p_{i,i-j}} \tag{2.64}
$$

where here $\eta_i(s) = E[e^{-sS_i}] = \alpha_i(\alpha_i + s)^{-1}$. To specialize this to a repair model, introduce a binomial distribution for successful repairs:

$$
\alpha_i = \lambda(n - i) + \mu
$$

$$
p_{i,i+1} = \lambda(n - i)\alpha_i^{-1} \tag{2.65}
$$

$$
p_{i,i-j} = \mu\alpha_i^{-1} \binom{i}{j} \rho^j (1 - \rho)^{i-j}, \qquad j = 0, 1, 2, \ldots, i
$$

where ρ represents the probability that an individual down machine is indeed repaired just after inspection (neglect the duration of repair times). The binomial model assumes that repair success is independent across machines, which may be inappropriate in case similar causes give rise to the failures. Differentiation or direct expectations yield moments of U_i and eventually of T_{0i}.

2.7 DIFFUSION AND FLUID APPROXIMATION

While classical discrete state space Markov process ideas can often be used to model some quite interesting situations, the analytical results obtained frequently emerge only in terms of incomprehensible transforms, or in other somewhat obscure form. Not infrequently the difficulty that induces complexity can be traced back to the influence of boundaries upon the process transitions. If one examines a rather heavily loaded or congested service system, however, it is apparent, first, that the state changes may appear almost negligibly small relative to the system state magnitude (e.g. length of queue) itself, and, second, that annoying boundaries, particularly that at zero, are visited infrequently—although their influence may still be crucial. These remarks hold true not only for simple one-dimensional processes, such as those used to describe congestion at a single servicing facility, but also for much more complex situations involving the interaction of several servicing processes.

An attractive approach to problems involving many customer arrivals occurring rapidly and generating considerable queueing is, then, to treat them by the method of *diffusion process* approximation. For details concerning the rigorous details of diffusion mathematics see Feller (1966); in brief summary recall that a diffusion is a possibly vector-valued Markov process on the real numbers that typically moves continuously, governed by a drift (infinitesimal mean) and a diffusion (infinitesimal variance) parameter.

This approximation has been employed by Kobayashi (1983) for certain cyclic networks of queues such as are encountered in multiprogramming computer systems. (See also Gaver and Shedler, 1973, the important work of Reiman, 1982, and particularly Newell, 1979 and also McNeil and Schach, 1973.) In this section the use of diffusion will be briefly illustrated, and some experience with the results will be recounted.

Situation (waiting time or backlog at one server): suppose a single servicing facility is confronted by random arrivals that bring with them contributions to work load, expressed as required processing times. If the facility processes them in order of arrival, what is the backlog at time t? The facility is heavily loaded, so that it is seldom idle. It never turns away customers, i.e. infinite buffering is possible, nor do long delays discourage those waiting, causing defections or balking.

Model: assume that arrivals occur in a Poisson (λ) process, and that the generic processing time S has df $G_S(y)$; successive processing times are IID. This model has been studied by Takaçs (1962) who derived an integro–differential equation for the df of backlog or virtual waiting time $W(t)$. Although the formal solution of that equation can be obtained, it is in a somewhat complicated form, not conducive to immediate insights. It is tempting to take an alternative, somewhat heuristic approach. Intuitively, if $\rho = \lambda E[S]$ (= expected total load increment per unit time) > 1 (= processing or output rate per unit time), the backlog grows at rate $\rho - 1 > 0$. Furthermore, the backlog process, $W(t)$, "ignores the boundary" at $W = 0$ after a time, and eventually $W(t)$ appears approximately normal/Gaussian over an interval $(t, t + \Delta)$, with mean $(\rho - 1)\Delta$, and variance $\lambda E[S^2]\Delta$. Importantly, also, the process seems to grow by accumulating independent, nearly Gaussian, increments.

If $\rho < 1$, some difficulties occur because $W = 0$ is an impermeable reflecting boundary, but the basic scenario is still the same: if ρ is close to unity, $W(t)$ moves in nearly Gaussian increments, but occasionally interacts with the boundary at zero. Question: what is the long-run behavior of the delay in such a process?

The Takács (or forward Kolmogorov) equation for the "exact" process is

$$-\frac{\partial F}{\partial x} + \frac{\partial F}{\partial t} = -\lambda F + \lambda \int_0^x F(x - y, t)G_S(dy) \qquad (2.66)$$

where $F(x, t)$ is the df of $W(t)$; initial and boundary conditions are necessary but are suppressed.

If $F(x, t)$ is only appreciable when x is large, and if the magnitude of a typical S is also small compared to x then it becomes plausible to Taylor-expand the $F(x - y, t)$ term to three terms and integrate; the result is

$$\cdot \frac{\partial \tilde{F}}{\partial t} = (1 - \lambda E[S]) \frac{\partial \tilde{F}}{\partial x} + \frac{\lambda E[S^2]}{2} \frac{\partial^2 \tilde{F}}{\partial x^2} \qquad (2.67)$$

which is the well known forward partial differential equation for Brownian motion with drift. Impose the boundary condition that $\tilde{F}(x, t) = 0$ for $x < 0$, and the equation for $\tilde{F}$, an approximation to F, the "true" distribution of $W(t)$, emerges; the latter equation can be explicitly solved in terms of error functions and exponentials (see Newell, 1971) for all t; i.e. the transient solution is actually readily available. With some further effort one can impose an upper boundary at $\bar{x} > 0$ to represent a finite buffer size. The approximate steady state ($t \to \infty$) distribution turns out to be

$$\tilde{F}(x) = \exp\left[\frac{-2(1 - \rho)}{\lambda E[S^2]} x\right] \qquad \rho < 1 \qquad (2.68)$$

which often is, for large ρ, usefully close to the behavior of $F(x)$, the long-run solution to the Takács equation.

Note that the parameters of the above differential equation can be obtained by equating infinitesimal mean and variance of the assumed Poisson arrival process to the corresponding quantities for the diffusion. It is interesting that, when available, a *martingale* approach to the problem, of

Gaver and Shedler (1973), produces a *different* exponent that yields better approximations for moderate traffic intensities, especially if the service time distribution is very long tailed (more skewed than the exponential).

We turn now to a more complex example, involving the interaction of two traffic streams.

Situation: At a node of communication network there are a total of $c + v$ channels (servers); voice messages are exclusively assigned to the v channels, and data messages are assigned to the c channels, *but* may also utilize any unused voice channel capacity under the stipulation that voice has preemptive priority and may displace any data encroaching on its (v channel) territory. Question: what is the nature of the delay experienced by the data, which is allowed to queue up in a buffer indefinitely?

Model (Markov assumptions): voice traffic is Poisson (λ) with expon(μ) service times; voice is a loss system, so immediately the steady-state voice loss rate can be calculated using the Erlang-B formula. The data, however, operate in a random service environment modified by voice needs. Data arrive in an independent Poisson (δ) process, with independent expon(η) service times. Typically $\delta \gg \lambda$, and $\eta \gg \mu$. Data, with state variable $X(t)$, are an M/M/s system where $s = c + v - V(t)$, $V(t)$ being the number of voice messages in service. Clearly $\{X(t), V(t)\}$ is a bivariate Markov process, but one difficult to analyze exactly; see references in Gaver and Lehoczky (1981) for other approaches to the analysis.

Now typically η/μ ($\simeq$ data arrival rate per voice service time) is very large, possibly 10^4. Furthermore, often $\rho_d \equiv \delta/\eta c$, so some voice channel usage by data is necessary in order that all data be handled and there is not an evergrowing queue. The appropriate traffic intensity parameter for the system is seen to be

$$\rho = [\rho_d + \rho_v(1 - q)](c + v)^{-1} \qquad (2.69)$$

where

$$\rho_v = \lambda/\mu,$$

and

$$q = \frac{\rho_v^v/v!}{\sum_{j=0}^{v} \rho_v^j/j!}$$

is the probability that the voice system rejects an arriving voice message. If ρ becomes large under such circumstances, Gaver and Lehoczky (1981) show that $X(t)$ behaves like a Wiener process with reflecting boundary, precisely as was mentioned in the previous example. Actually Gaver and Lehoczky (1981) assume that data input acts as a *fluid*, with no variability.

By further, more intricate, methods involving the convergence of semi-groups of operators developed by Burman (1979), it is shown in Gaver and Lehoczky (1981) that the long-run distribution of $X(t)$ is approximately exponential with mean

$$\rho_d + \frac{\dfrac{\eta}{\mu\rho_r} \displaystyle\sum_{i=0}^{v-1} (T_i^2/\pi_i)}{(c + v)(1 - \rho)}$$

where   (2.70)

$$\pi_i = \rho_v^i/(i!)\Big/\left[\sum_{i=0}^{v} \rho_v^i/(i!)\right]$$

and

$$T_k = \sum_{i=0}^{k} \pi_i(i - \rho_v(1 - q))$$

Numerical work indicates that the diffusion approximation is reasonably accurate if the traffic intensity is quite high, say of $\rho \geq 0.95$; otherwise, for smaller s, the accuracy is not as high. It would not be surprising if a refined method for fitting drift and diffusion coefficients would lead to improved results. Finally, the difference between the refined treatment of basic data as a Poisson, and the simpler treatment by a fluid (yielding a model that can be solved exactly), resides merely in the addition of the term ρ_d in the numerator of the expression for the mean.

The diffusion approximation can be utilized to evaluate another interesting measure of system effectiveness, namely the expected total waiting time, in job or data-packet hours, expended during a busy period for data traffic that starts with x present.

Model: let $A(x)$ be the expected total waiting time during a busy period when the initial number of jobs is $X(0) = x > 0$. Condition on process change, $Z(\Delta)$, during the initial short time period $(0, \Delta)$ to get

$$E\left\{\int_0^\infty X(t')\,dt' \,\big|\, X(0) = x, X(\Delta) = x + Z(\Delta)\right\}$$

$$\equiv A(x; Z(\Delta)) = x\Delta + A(x + Z(\Delta)) + o(\Delta) \qquad (2.71)$$

Now Taylor-expand and remove the condition on $Z(\Delta)$:

$$A(x) = x\Delta + A(x) + A_x(x)\mu(x)\Delta + A_{xx}(x) \cdot \frac{\sigma^2(x)}{2}\Delta + o(\Delta) \qquad (2.72)$$

or, collecting terms in Δ and letting $\Delta \rightarrow 0$,

$$0 = x + \mu(x)A_x(x) + 1/2\sigma^2(x)A_{xx}(x) \qquad (2.73)$$

to be solved subject to $A(0) = 0$; clearly restrictions on $\mu(x)$ are necessary in order that $A(x)$ be finite.

If $\mu(x) = \mu < 0$, and $\sigma^2(x) = \sigma^2$, both independently of x, the equation can be solved directly to give

$$A(x) = \frac{1}{2\mu}x^2 + \frac{\sigma^2}{2\mu^2}x \qquad (2.74)$$

where $\mu < 0$ when traffic intensity $\rho < 1$. A similar expression for compound Poisson (λ) inputs is

$$A(x) = \frac{x^2}{2(1-\rho)} + \frac{xE[A^2]}{2(1-\rho)^2} \qquad \rho < 1 \qquad (2.75)$$

where A represents the number of items (packets) arising at a single request; $\rho = \lambda E[A]$. The moral is that *variability* (measured by σ^2 or $E[A^2]$) can greatly increase *expected* total waiting time, particularly when system loading is high (ρ close to unity).

For application of this "area under a random path" to discussing total wait during a road traffic jam, see Gaver (1969). See also McNeil (1970) for generalizations. The same backward argument is well-adapted also to studying problems of optimum investment decisions.

2.8 RENEWAL-THEORETIC MODELING

Ideas of renewal theory and recurrent events are extremely useful for many purposes in stochastic modeling. Recognition of the occurrence of one or more recurrent events or "renewal points" in the development of a model process points the way to writing down simple forward or backward type equations for probabilities or expectations. Frequently analytical information can be extracted from such equations, particularly that relevant to long-time or other asymptotic process behavior. If more information is desired it can be obtained by use of transform techniques, by numerical computation, or by Monte Carlo simulation.

Mathematical definitions and properties of renewal processes are well presented by Feller (1966), Karlin and Taylor (1975), and Cox (1962), among others. There follow a few situations and suggested models based on renewal theory that illustrate the basic notions. We also comment on the relevance of the models and results obtained to real situations.

Situation: a machine, e.g. a computer system or component thereof, or human operator, etc., operates properly for a period of time, fails, is restored (or restores itself) to service and operates properly again for a different time, fails again, and so on. Questions: how many failures are likely to occur in a given fixed period of time, say a year? The answer to such a question will help to guide decisions concerning logistics (necessary spare parts) and employment of repair personnel. How long a time will elapse until the kth failure? Suppose the times to restore failures vary; what is the likelihood that a "chance" user of the system will find the machine down for repair when he or she needs it, and how long will the wait to service restoration last? These are only a few of the many questions that might be asked.

Model: the classical renewal theory model for the situation described postulates that times between successive failures, $\{X_i, i = 1, 2, \ldots\}$ are IID positive RVs with distribution $F_X(x)$. That is X_1 is the time until the first event (here failure), and X_i is the elapsed time between the $(i - 1)$st and the ith event. X_i can be either a discrete RV, so failures occur at regular intervals, say hourly, a continuous RV or a mixture. For the moment assume repairs to take a negligible time.

Mathematical results for this model are simplest and nicest when the IID assumption is fully exploited. Under that assumption (and even more generally) the counting process, $N(t)$, giving the number of renewal events (failures in time t) has probability distribution

$$P\{N(t) = n\} = F_X^{n^*}(t) - F_X^{(n+1)^*}(t) \tag{2.76}$$

where * refers to convolution. For long time ($t \to \infty$) and under suitable mathematical restrictions

$$M(t) \cong E[N(t)] \simeq \frac{t}{E[X]} \tag{2.77}$$

$$\mathrm{Var}[N(t)] \simeq \frac{\mathrm{Var}[X]}{(E[X])^3} t$$

and $N(t)\sim$normal, with the above parameters. Of course exact analytical solutions can be obtained if sympathetic distributional models are assumed: taking $X\sim$expon yields the Poisson distribution, and $X\sim$gamma or Erlang also produces rather neat closed-form solutions. Any discrete-time distribution for X can be numerically convolved conveniently using APL. This helps to answer questions about the number of events in $(0, t)$, *provided* the IID assumption is palatable. If finite-time results are needed, resort can be made to numerical summation, using a discrete time model, to approximation by a standard, tractable, distribution such as the gamma followed by transform inversion, or by simulation.

In what follows we illustrate the diverse utility of backward conditioning arguments, leading to renewal integral equations.

Situation (incorrect repair possibly due to human error): each time a repair is made there is the chance that it will be incorrect, and that the subsequent time to failure will be short. Suppose that incorrect repairs tend to bunch together in runs; describe the number of failures that occur over a (long) time period.

Notice that a similar situation describes a clustering scheme of arrivals to a repair facility or a communications center.

Model: assume that the generic time to failure of the system is X when repair is made properly, and X' when repair is incorrect; the intermingled sequence of X and X' quantities are conditionally independent. Furthermore, let the probability of a correct repair at failure n be α, $0 \le \alpha \le 1$, if the repair at the time of previous failure $n - 1$ was correct, and $\bar{\beta}, = 1 - \beta$, $0 \le \beta \le 1$ ($\alpha \ne \beta$), if it was incorrect; the sequence of correct and incorrect repairs is thus modeled as a stationary ergodic Markov chain. This, of course, does not represent systematic improvements in repair capability, although a transient chain could serve for that purpose.

Let $M(t)$ ($M'(t)$) denote the mean or expected number of repairs in t, *given* that the first repair was correct (incorrect); think of the first repair (manufacture) as occurring at $t = 0$. Argue that

$$M(t) = \begin{cases} 1 & \text{if } X > t; \\ 1 + M(t - X) & \text{if } X \le t \text{ and the 2nd repair is correct;} \\ 1 + M'(t - X) & \text{if } X \le t \text{ and the 2nd repair is incorrect.} \end{cases} \tag{2.78}$$

Likewise,

$$M'(t) = \begin{cases} 1 & \text{if } X' > t; \\ 1 + M(t - X') & \text{if } X' \le t \text{ and the 2nd repair is correct;} \\ 1 + M'(t - X') & \text{if } X' \le t \text{ and the 2nd repair is incorrect.} \end{cases} \tag{2.79}$$

Now if the various conditions are removed then according to the model there results the two linked convolution integral equations:

$$M(t) = 1 + \alpha \int_0^t M(t - x)F_X(\mathrm{d}x) + (1 - \alpha)\int_0^t M'(t - x)F_{X'}(\mathrm{d}x)$$

$$\tag{2.80}$$

32   *Donald P. Gaver*

and

$$M'(t) = 1 + \beta \int_0^0 M'(t - x)F_{X'}(\mathrm{d}x) + (1 - \beta) \int_0^t M(t - x)F_X(\mathrm{d}x)$$

(2.81)

In turn, these two equations are susceptible to transforming: multiply by e^{-st} and integrate to get

$$\hat{M}(s) = \frac{1}{s} + \alpha\hat{M}(s)\hat{F}_X(s) + \bar{\alpha}\hat{M}'(s)\hat{F}_{X'}(s)$$

and

$$\hat{M}'(s) = \frac{1}{s} + \beta\hat{M}'(s)\hat{F}_{X'}(s) + \bar{\beta}\hat{M}(s)\hat{F}_X(s);$$

matrix notation is natural here, especially if more than two repair states are used. If one then solves and collects terms to order $(1/s)^2$ as $s \to 0$, Tauberian theorems, cf. Feller (1966), show that for large t

$$M(t) \approx M'(t) \sim \frac{\bar{\alpha} + \bar{\beta}}{\bar{\alpha}E[X'] + \bar{\beta}E[X]} t$$

(2.82)

and a little reflection shows that this is entirely sensible. In similar ways variances can be written down, and an approximate normal/Gaussian distribution for total failures may be derived. The model can also be extended to account for the existence of nonzero repair times, and total *availability* studied.

Here is another, somewhat more complicated, application of renewal theory ideas, now to an inspected system problem.

Situation (standby system availability): a system, such as an emergency electric power source, is usually in a quiescent or cold standby status, but occasionally is called upon to fill a function (e.g. generate power). Various systematic plans might be devised for assuring reasonably high system availability, or probability of satisfying a demand when one occurs. One such plan is to inspect infrequently so long as no failure is detected between inspections, and otherwise inspect more frequently until evidence of need seems gone. Problem: develop a model to evaluate such an inspection scheme.

Model: let the inspection plan be to inspect at *long* intervals, $\{L_i, i = 1, 2, \ldots\}$ until such time as an inspection reveals a failure, and then switch to *short* intervals $\{S_i, i = 1, 2, \ldots\}$, continuing until there has been a run of r (10, say) failure-free short-interval inspections, at which time switch back to long intervals; continue indefinitely. A measure of effectiveness is the long-run point availability of the system, i.e. the probability that the system is failure-free on the occasion of a demand.

To evaluate such a rule, allow the L_i and S_i sequences to be IID and mutually independent, with dfs $F_L(x)$ and $F_S(x)$ respectively; if desired these latter can be specialized to concentrate at fixed values (e.g. 14 days and 1 day, respectively). Furthermore, let λ be the failure rate of a system failing at exponentially distributed intervals even when "cold". In order to analyze the system by renewal theory it is worthwhile to look at the periods during which inspections are infrequent, called *L eras*, and those alternating with them, during which inspections occur frequently, called *S eras*. Note that a demand can occur during either type of era; L and S eras constitute an *alternating* renewal process.

The analysis involves the following components.

2.8.1 Distribution (Density) of *L*-era Duration

Suppose an inspection has just been completed at $t = 0$ and nothing amiss has been detected. Furthermore, suppose that this inspection marked the end of the previous S era, so an L era is just beginning. Let $a_L(\mathrm{d}t)$ be the probability that the present L era will last for time $(t, t + \mathrm{d}t)$, or, loosely, until exactly t. One can now write down a renewal equation for $a_L(\mathrm{d}t)$:

$$a_L(\mathrm{d}t) = (1 - e^{-\lambda t})F_L(\mathrm{d}t) + \int_0^t e^{-\lambda t'}F_L(\mathrm{d}t')a_L(\mathrm{d}t - t'); \qquad (2.83)$$

the first term on the right-hand side means that the L era terminates with the *first* inspection, meaning that the unit has failed before L_1. The second term represents survival through the first inspection at which time the process renews itself or starts anew; final failure occurs at time $t - t'$ thereafter. Failure or no failure at first inspection are mutually exclusive and exhaustive events, and so the result is a renewal equation for $a_L(\mathrm{d}t)$. Introduce transforms to find

$$a_L(s) \equiv E[e^{-sL}] = \frac{\hat{F}_L(s) - \hat{F}_L(\lambda + s)}{1 - \hat{F}_L(\lambda + s)} \qquad (2.84)$$

The mean of the L-era duration, denoted by $\underline{L}$, is

$$E[\underline{L}] = \frac{E[L]}{1 - \hat{F}_L(\lambda)} \qquad (2.85)$$

2.8.2   Distribution of *S* era Duration

If an inspection has just revealed a failure, an *S* era begins immediately (take inspection and repair to be instantaneous for the moment). Let $a_S(\mathrm{d}t)$ be the probability that an *S* era lasts for time *t*. Then the following renewal equation may be written:

$$a_S(\mathrm{d}t) = \mathrm{e}^{-\lambda t} F_S^{r*}(\mathrm{d}t) + \int_0^t h(\mathrm{d}t') a_S(\mathrm{d}t - t'); \qquad (2.86)$$

the auxiliary function *h* represents the probability that an inspection reveals a failure before the termination of the *S* era in progress; this causes the frequent inspection to start over, i.e. starts the *S* era afresh. In terms of transforms, $\underline{S}$ denoting an *S* era duration,

$$\hat{a}_s(s) \equiv E[\mathrm{e}^{-s\underline{S}}] = \frac{(\hat{F}_S(s + \lambda))^r}{1 - \hat{h}(s)}, \qquad (2.87)$$

and

$$\hat{h}(s) = [\hat{F}_S(s) - \hat{F}_S(\lambda + s)] \cdot \{1 + \hat{F}_S(s + \lambda) + (\hat{F}_S(s + \lambda))^2 + \ldots$$
$$+ (\hat{F}_S(s + \lambda))^{r-1}\} \qquad (2.88)$$
$$= \frac{\hat{F}_S(s) - \hat{F}_S(s + \lambda)}{1 - \hat{F}_S(s + \lambda)} \cdot [1 - (\hat{F}_S(s + \lambda))^r]$$

The latter transform expresses the probability that the necessary run of *r* successes is interrupted by a failure, and must begin again. From the transform comes the expected length of an *S* era:

$$E[\underline{S}] = E[S]\left\{\frac{(\hat{F}_S(\lambda))^{-r} - 1}{1 - \hat{F}_s(\lambda)}\right\} \qquad (2.89)$$

2.8.3   Probability that System is Available During an *L* era

The probability $A_L(t)$ that the system is up at time *t* after the beginning of an *L* era is simply $\mathrm{e}^{-\lambda t}$, the probability of no failure, and the transform is

$$\int_0^\infty \mathrm{e}^{-st} A_L(t)\,\mathrm{d}t \equiv \hat{A}_L(s) = (\lambda + s)^{-1} \qquad (2.90)$$

2.8.4   Probability that System is Available During an *S* era

If $A_S(t)$ is the probability that the system is up after an *S* era has progressed for time t, by renewal

$$A_S(t) = e^{-\lambda t}\bar{F}_S^{r^*}(t) + \int_0^t h(dt')A_S(t - t') \qquad (2.91)$$

where h has appeared before, under Section 2.8.2. It follows that

$$\hat{A}_S(s) = \frac{1 - (\hat{F}_S(s + \lambda))^r}{(s + \lambda)[1 - \hat{h}(s)]} \qquad (2.92)$$

2.8.5 Overall Availability at t

Let $A(t)$ denote the availability of the system at t. Again by backward renewal argument, and starting at the beginning of an L era,

$$A(t) = A_L(t) + \int_0^t A_S(t - t')a_L(dt') + \int_0^t A(t - t')a_L^* a_S(dt'); \qquad (2.93)$$

transforms then give

$$\hat{A}(s) = \frac{\hat{A}_L(s) + \hat{A}_S(s)\hat{a}_L(s)}{1 - \hat{a}_L(s)\hat{a}_S(s)}; \qquad (2.94)$$

a Tauberian theorem now shows that

$$\lim_{t \to \infty} A(t) = \frac{\hat{A}_L(0) + \hat{A}_S(0)}{E[\underline{L}] + E[\underline{S}]},$$

so the long-run point availability is

$$A(\infty) = (\lambda^{-1})\left\{\frac{(\hat{F}_S(\lambda))^{-r}}{E[L](1 - \hat{F}_L(\lambda))^{-1} + E[S]((\hat{F}_S(\lambda))^{-r} - 1)(1 - \hat{F}_S(\lambda))^{-1}}\right\} \qquad (2.95)$$

The expression for $A(\infty)$ is easily evaluated numerically if expressions for L-interval and S-interval transforms are obtainable. It is sometimes useful to interpret

$$\hat{A}(s) \cdot s = \int_0^\infty A(t)e^{-st}s\, dt \qquad (2.96)$$

as the availability of the system *upon demand*, the demand now occurring at a random time D, D having the exponential distribution with mean s^{-1}. In this case the initial conditions matter (they do not, in the long run), and a different availability figure is obtained depending upon whether the system is initially in an L era or an S era.

3 Additional modeling topics

In this chapter are outlines of certain modeling topics that formed the basis for cooperative research at ULB.

3.1 DISTRIBUTIONAL SCULPTURING OR INVERSE MODIFICATION

Standard distributions such as the exponential or gamma in particular, and also the normal, log-normal, and many others may reasonably and conveniently serve as components of a stochastic model or be used to summarize data distributions. For instance, interarrival times to service systems may appear approximately exponential, service times nearly gamma or log-normal, and so on. On the other hand, a systematic departure from such a standard may be revealed by analyzing actual data. It is then frequently possible to alter conventional distributions, or, equivalently, transform the RV in simple ways in order to represent empirical reality more closely. Here are two conventional examples; there then follow some more general procedures.

Example 1: the *Weibull* distribution is often utilized to represent times to failure of components or times between system demands. Let T be a RV (time, for instance), then T is distributed in a Weibull manner if

$$F_T(t; \alpha, \beta) = P\{T \le t\} = \begin{cases} 1 - e^{-\alpha t^\beta} & t > 0, \alpha > 0, \beta > 0 \\ 0 & t < 0 \end{cases} \quad (3.1)$$

The Weibull density is

$$f_T(t; \alpha, \beta) = e^{-\alpha t^\beta} \alpha \beta t^{\beta-1} \quad (3.2)$$

36

Furthermore, the Weibull hazard or failure rate at age t is

$$\frac{1}{dt} P\{T \in (dt) | T > t\} \equiv k_T(t; \alpha, \beta)$$

$$= \frac{f_T(t; \alpha, \beta)}{1 - F_T(t; \alpha, \beta)} = \alpha\beta t^{\beta-1} \qquad (3.3)$$

The latter formula and also (3.2) imply that if $\beta = 1$ the Weibull is actually an exponential; for this case "age" or "time in service" as measured by t does not influence the probability of failure in the next short time interval $(t, t + dt)$. On the other hand, $\beta > 1$ implies that the rate of failure having survived to t—the hazard—increases with age, while, $\beta < 1$ implies that the hazard decreases with age. If $\beta < 1$ the Weibull right tail is longer than that of the exponential (extremely positively skewed), while if $\beta > 1$ the positive skewness is less pronounced. The Weibull RV, T, is actually only an exponential RV, X, transformed or disguised, for if

$$P\{T \le t\} = 1 - e^{-\alpha t^\beta} = P\{T^\beta \le t^\beta\}, \qquad (3.4)$$

then

$$P\{T^\beta \le x\} = 1 - e^{-\alpha x}$$

Consequently, $T^\beta = X$, an exponential RV, and $T = X^{1/\beta}$ is a representation of a Weibull in terms of an exponential. Supposing that one wishes to simulate a Weibull (α, β) RV, then one merely simulates an expon (1) RV and raises it to the $(1/\beta)$th power, after multiplying by α^{-1}. Since the power transformation is monotonic increasing, the *quantiles* of Weibull and exponential are related by

$$t(p) = (x(p))^{1/\beta} \qquad (3.5)$$

Example 2: the *log-normal* distribution is a favorite model for system repair times (see Kline and Almog, 1980); it has many other uses, and even some rational persuasions for its apparent resemblance to empirical distributions (see Aitchison and Brown, 1947). Say that X is log-normal if $\ln X = Y$ is normal (it does not matter what the *base* of the logs is!); specifically $Y \sim N(\mu, \sigma^2)$. Here are some properties:

$m_k = e^{k\mu + k\sigma^2/2} = E[X^k]$

$m_1 = E[X] = e^{\mu + \sigma^2/2}$, $\text{Var}[X] = (E[X])^2 (e^{\sigma^2} - 1) \equiv (E[X])^2 \eta^2$

$\gamma_1(X) = \text{skew}[X] = \eta^3 + 3\eta$ $\qquad (3.6)$

$\gamma_2(X) = \text{kur}[X] = \eta^8 + 6\eta^6 + 15\eta^4 + 16\eta^2$

$\text{Median}[X] = e^\mu$, $\quad \text{Mode}[X] = e^{\mu - \sigma^2}$

Supposing that one wishes to simulate a log-normal RV, one simply simulates an $N(\mu, \sigma^2)$ RV, Y and exponentiates $X = e^Y$.

The above two familiar examples illustrate the formation of new and useful random variables and distributions by simple transformation. An intuitively appealing way of looking at certain transformations is in terms of modifications to familiar RVs or quantiles by convenient *shaping factors*. This process will be called *distributional sculpturing*: let X be a basic RV, for example an exponential, normal or log-normal. Define

$$Y = X s(X) \qquad (3.7)$$

where the *shaping function* $s(X)$ is designed to conveniently convert the basic RV X into a shaped version, Y, having the desired distributional properties. Some examples of the properties often desired in practice, along with suitable—but definitely not unique—shaping functions, now follow.

3.1.1 Skewness-Producing Shapers

Examples: X is a positive basic RV, e.g. exponential.

$$\text{(i) } s(X) = 1 + AX^l, \qquad A > 0, l > 0 \qquad (3.8)$$

$$\text{(ii) } s(X) = e^{AX^l}$$

Effect:

$$Y = X s(X) \approx \begin{cases} X & \text{if } X \text{ "small"} \\ \text{(i) } X + AX^{l+1} \\ \text{(ii) } Xe^{AX^l} \end{cases} \gg X \Big\} \text{ much greater than } X \text{ if } X \text{ "large"}$$

The particular shaping functions (i) and (ii) both leave small values of X unchanged, but considerably expand large values, thus transforming the distribution of X, e.g. the exponential, to one that is nearly exponential near the origin, but having a relatively long or stretched right tail. For example, take the shaping function (*i*) with $l = 1$ and apply to X exponential. The shaped distribution becomes

$$F_Y(y; A) = 1 - \exp\left[-\left(\frac{\sqrt{1 + 4Ay} - 1}{2A}\right)\right] \qquad (3.9)$$

Examination shows that for small y (Taylor series) the distribution of Y is nearly unit exponential, while for large Y it resembles a Weibull with shape parameter $\beta = 1/2$. Shaping function (ii) has an even more pronounced effect on the right tail. Note that both shaping functions (i) and (ii) yield monotonic increasing transformations from $X \rightarrow Y$, and that given by (i) is sometimes explicitly algebraically invertible (solve a quadratic equation when $l = 1$, cubic when $l = 2$, quartic when $l = 3$), while that of (ii) is not. Also note that useful transformations result when the parameter $A < 0$: this actually may result in right tail truncation (severe shortening); such transformations are no longer monotonic.

Moments: The moments of shaped or sculptured RVs can sometimes be conveniently calculated. For the present representations:

(i) $Y = X(1 + AX^l)$

$m_1(Y) = m_1(X) + Am_{1+l}(X)$

$m_2(Y) = E[(X(1 + AX^l))^2]$

$\quad = m_2(X) + 2Am_{1+l}(X) + A^2 m_{2+2l}(X)$ (3.10)

$\mathrm{Var}[Y] = \mathrm{Var}[X] + 2A\,\mathrm{Cov}[X, X^{l+1}] + A^2\,\mathrm{Var}[X^{l+1}]$

Also, to indicate the dependence between the stretched RV Y and the basic RV X,

$$m_k(Y - X) = A^k m_1(X^{(l+1)k}) = A^k m_{(l+1)k}(X),$$

so

$$\mathrm{Var}[Y - X] = A^2\,\mathrm{Var}[X^{l+1}] \quad (3.11)$$

and

$$\mathrm{Cov}[Y, X] = \mathrm{Var}[X] + A\,\mathrm{Cov}[X, X^{l+1}] \quad (3.12)$$

The quantiles are directly and simply related by

$$Y(p) = x(p)(1 + A(x(p))^l), \quad 0 \le p \le 1 \quad (3.13)$$

(ii) $Y = Xe^{AX^l}$, but for the present consider only $l = 1$. Then the kth moment is expressible in terms of the derivatives of the Laplace transform of X. Note that the kth moment does not necessarily exist for all basic distributions. When it does,

$$m_k(Y) = (-1)^k \frac{d^k}{ds^k} E[e^{-sX}]|s = -kA \quad (3.14)$$

For example, if $X \sim$ expon (1),

$$m_k(Y) = \frac{k!}{(1 - kA)^{k+1}} \quad \text{if} \quad kA < 1 \quad (3.15)$$

otherwise the moment does not exist because the right tail of the distribution is too long.

3.1.2 Symmetric Stretch-Producing Shapers

Example: Z is a RV symmetrically distributed around zero; e.g., $Z \sim N(0, \sigma^2)$. Here are some useful shapers:

(i) $s(Z) = 1 + hZ^2, \quad h > 0$; and

(ii) $s(Z) = e^{hz^2}$ (due to J. W. Tukey).

Again (i) and (ii) imply that $Y = Zs(Z)$ resembles Z for small Z, but lengthens the tails of the distribution for large Z.

Example 1: stretched log-normal variables may be suggested for modeling repair or service times if data analysis indicates that the logarithms of observed times are symmetric but not nearly normal, having symmetrically too-long tails. Then it may be convenient to use the representation

$$Y = e^X$$

$$X = \mu + cZs(Z) \tag{3.16}$$

where μ is the center (corresponds to mean) of the logged observations, and the scale constant c is the standard deviation (spread parameter) of the variable $Zs(Z)$, replacing σ in the ordinary log-normal formula.

Moments: The moments of the shaped distribution are, for (i), representable in terms of those for a basic Z.

(i) $\quad Y = Z(1 + hZ^2)$

$$m_1(Y) = m_1(Z) + hm_3(Z)$$

$$= 0 \ (Z \text{ symmetrical around zero}) \tag{3.17}$$

$$m_2(Y) = m_2(Z) + 2hm_4(Z) + h^2m_6(Z)$$

$$m_4(Y) = m_4(Z) + 4hm_6(Z) + 6h^2m_8(Z) + 4h^3m_{10}(Z) + h^4m_{12}(Z)$$

(ii) $Y = Ze^{hZ^2}$; consider only $Z \sim N(0, \sigma^2)$. Calculate as a preliminary

$$\int_{-\infty}^{\infty} e^{hz^2} e^{-1/2z^2/\sigma^2} \frac{1}{\sqrt{2\pi}\sigma} = (1 - 2h\sigma^2)^{-1/2} \quad h < 2\sigma^2 \tag{3.18}$$

Then differentiate repeatedly with respect to h to obtain the even moments (the odd moments equal zero):

$$\text{Var}[Y] = m_2(Y) = \frac{\sigma^2}{(1 - 4h\sigma^2)^{3/2}} \quad h < 1/4\sigma^2$$

$$m_4(Y) = \frac{3\sigma^4}{(1 - 8h\sigma^2)^{5/2}} \quad h < 1/8\sigma^2 \tag{3.19}$$

Hence kurtosis is

$$\gamma_2 = \frac{m_4(Y)}{(m_2(Y))^2} = \frac{3(1 - 4h\sigma^2)^3}{(1 - 8h\sigma^2)^{5/2}} - 3 \qquad (3.20)$$

Tails are so extended that the kurtosis becomes very large in the case of (i), and actually infinite for rather small values of h in (ii); the variance remains finite for slightly larger values. Nevertheless, the central part of the Y-distribution remains remarkably close to the normal form when Z is itself normal.

Both forms (i) and (ii) can be induced to fit the inverse distribution (per cent points) of the Student's t distribution fairly satisfactorily.

3.1.3   Left Tail Enhancement Shapers

Example: X is a positive RV, e.g. exponential.

$$\text{(i)} \ \ s(X) = \frac{\alpha X^l}{1 + \alpha X^l} \qquad l > 0, \, \alpha > 0$$

$$\text{(ii)} \ \ s(X) = 1 - e^{-\alpha X^l} \qquad (3.21)$$

$$\text{(iii)} \ \ s(X) = e^{-\alpha/X^l}$$

This shaping function tends to concentrate probability near zero, leaving the distribution's shape unchanged for large X:

$$Y = Xs(X) \approx \begin{cases} X & \text{if } X \text{ large} \\ \text{(i)} \\ \qquad\qquad\Big\} \ \alpha X^{l+1} \ll X, X \text{ small} \\ \text{(ii)} \end{cases} \qquad (3.22)$$

Moments are generally impossible to find in usable form, although in the case of (ii), $l = 1$, explicit results can be found in terms of Laplace transforms: suppose $X \sim \text{expon}(1)$, then under (ii)

$$E[Y] = E[X(1 - e^{-\alpha X})] = 1 - \frac{1}{(1 + \alpha)^2} ; \qquad (3.23)$$

the mean of Y is influenced very little when α is large, but small values of X are made even smaller; for the expon(1) X the quantiles are related as follows:

$$y(p) = x(p)[1 - e^{\alpha \ln(1-p)}], \qquad (3.24)$$

so for a fixed $p > 0$ large α forces $y(p)/x(p)$ to one. However, for a fixed α

and small p

$$\frac{y(p)}{x(p)} \simeq \alpha p \simeq 0 \tag{3.25}$$

showing that low quantiles for X transform into even smaller values for Y. Furthermore, it is easy to see that if $X_{(1;n)}$ is the minimum in a sample of n from $X \sim$ expon (1), and

$$Y_{(1;n)} = X_{(1;n)}[1 - e^{-\alpha X_{(1;n)}}]$$

then

$$\frac{E[Y_{(1;n)}]}{E[X_{(1;n)}]} \simeq \frac{\alpha}{n + \alpha} \to 0 \quad \text{as} \quad n \to \infty \tag{3.26}$$

All of this reinforces the image of the present shaper as forcing small X values to become smaller, leaving large values unchanged. Note that the result of the shaper (i), $l = 1$, can be explicitly inverted, giving the distribution

$$F_Y(y) = F_X\left[\frac{y(1 + \sqrt{1 + 4/\alpha y})}{2}\right] \tag{3.27}$$

from which the left tail enhancement property shows itself explicitly:

$$f_Y(y) = f_X\left[\frac{y(1 + \sqrt{1 + 4/\alpha y})}{2}\right] \cdot \left\{1 + \frac{y + (2/\alpha)}{\sqrt{y^2 + \left(\frac{4}{\alpha}\right)y}}\right\} \cdot \left(\frac{1}{2}\right) \tag{3.28}$$

$$\simeq \begin{cases} f_X\left(\dfrac{\sqrt{y}}{\sqrt{\alpha}}\right) \dfrac{1}{2\sqrt{\alpha}\sqrt{y}} & \text{for } y \text{ small} \\[2ex] f_X(y) & \text{for } y \text{ large} \end{cases}$$

Note that the *hazard* associated with the latter transformation is, when X is an expon (1) RV,

$$k_Y(y) = \frac{1}{2}\left\{1 + \frac{y + 2/\alpha}{\sqrt{y^2 + (4/\alpha)y}}\right\}$$

$$\simeq \begin{cases} 1/2\sqrt{\alpha y} & \text{for } y \text{ small} \\ 1 & \text{for } y \text{ large} \end{cases} \tag{3.29}$$

Thus the density of an exponential X remains, under this transformation, exponential-like in the right tail but approaches infinity near the origin. A distribution with such behavior may well be useful for modeling failure data exhibiting early failures ("infant mortality").

3.1.4 Right Tail Shortening Shaper

Example: X is a positive RV, e.g. an exponential. To shorten or truncate the right tail, consider

$$\text{(i)} \quad s(X) = \frac{1}{1 + \beta X^l}, \qquad \beta > 0, l > 0$$

$$\text{(ii)} \quad s(X) = e^{-\alpha X^l}, \qquad \alpha > 0, l > 0 \tag{3.30}$$

The shaper (i) with $l = 1$ is a right tail shortener or truncator; it provides a monotonic transformation to $Y \le \beta^{-1}$; small values of X are left nearly unchanged. The quantiles of Y are

$$y(p) = \frac{x(p)}{1 + \beta x(p)} \simeq \begin{cases} x(p) & \text{for } p \text{ small} \\ \\ \beta^{-1} & \text{for } p \text{ large if } x(p) \gg \beta^{-1} \end{cases} \tag{3.31}$$

Shaper (ii) is nonmonotonic: values of $Y = Xs(X)$ increase to $(\alpha l)^{-1}$, decreasing thereafter. The inverse of (i) yields

$$\begin{cases} F_Y(y) = F_X\left(\frac{y}{1 - \beta y}\right) & 0 \le y \le \beta^{-1} \tag{3.32} \end{cases}$$

with density

$$f_y(y) = f_X\left(\frac{y}{1 - \beta y}\right) \frac{1}{(1 - \beta y)^2}; \tag{3.33}$$

the latter approaches infinity at a rapid rate as y nears $1/\beta$.

Shaper (i) with $l = 1/2$ is a tail thinner; it can be inverted to give

$$F_y(y) = F_X\left[\left(\frac{\beta y + \sqrt{\beta^2 y^2 + 4y}}{2}\right)^2\right] \tag{3.34}$$

with density

$$f_Y(y) = f_X\left[\left(\frac{\beta y + \sqrt{\beta^2 y^2 + 4y}}{2}\right)^2\right] \left[\frac{1}{2}(\beta y + \sqrt{\beta^2 y^2 + 4y})\left(\beta + \frac{\beta^2 y + 2}{\sqrt{\beta^2 y^2 + 4y}}\right)\right]$$

$$\simeq \begin{cases} f_X(y) & \text{for } y \text{ small} \\[2mm] f_X(\beta^2 y^2) 2\beta^2 y & \text{for } y \text{ large;} \end{cases} \qquad (3.35)$$

if X is exponential, then the hazard rate of Y is initially flat, but eventually increases linearly with age, thus producing a plausible wearout model for equipment or biological organisms.

The above examples illustrate a few of the large number of possible ways in which the sculpturing idea can be used to extend the descriptive power of standard distributional models.

Problems of fitting such models to actual data are currently being addressed, as are applications to simulation and time series modeling. Use of shaped exponentials to evaluate scheduling procedures has been initiated in collaboration with J. P. Colard at ULB.

3.2 RESPONSE TIMES UNDER PROCESSOR SHARING

Consider a simple model of a time-sharing computer system in which N terminals access a single computer (server). Let repairman model conditions prevail, but *processor sharing* governs the service order: if j jobs ($0 \leq j \leq N$) are at the execution stage each receives one-jth of a time unit of service per time unit. In other words, if μ is the service rate (μ^{-1} is the expected service time under Markov–exponential assumptions), then the unconditional probability that any designated single job finishes in $(t, t + \Delta)$ is $\mu(\Delta/j) + o(\Delta)$.

Under such conditions it is possible to derive backward-type equations to describe the response of waiting time, R, of a newly initiated "tagged" job arriving from a previously idle terminal. In particular, consider

$$m_j(t) = E[R \mid W(R) = t, X(0) = j] \qquad (3.36)$$

where $W(R)$ is the amount of work or processing time required of the server, and $X(0)$ represents the number of jobs currently at the processor when the tagged job first arrives (including that job). Thus $m_j(t)$ is the expected response time, conditional on need and accompaniment. Additionally, introduce $r(j)$ as the fraction of a time quantum, Δ, actually available for job processing when j jobs are present at the computer; $r(j)$ represents one component of overhead, and may decrease as j increases. For short, let λ^j represent the rate of new arrivals when j are present at the server; under the stated model conditions $\lambda_j = \lambda(N - j)$. Let μ_j be the rate at which jobs accompanying the tagged job depart; under the stated conditions $\mu_j = \mu(j - 1)(r(j)\Delta/j) + o(\Delta)$. These assumptions, or those for a general birth-and-death process, lead by a backward-conditioning to this equation:

$$m_j(t) = \Delta + m_j\left(t - \left(\frac{r(j)}{j}\right)\Delta\right)[1 - (\lambda_j + \mu_j)\Delta] + \lambda_j\Delta m_{j+1}\left(t - \frac{r(j)\Delta}{j}\right)$$

$$+ \mu_j\Delta m_{j-1}\left(t - \frac{r(j)\Delta}{j}\right) + o(\Delta) \qquad (3.37)$$

Subtract $m_j(t - r(j)\Delta/j)$ from each side, divide by Δ, and let $\Delta \to 0$ to obtain

$$\left(\frac{r(j)}{j}\right)\frac{dm_j(t)}{dt} = 1 - (\lambda_j + \mu_j)m_j(t) + \lambda_j m_{j+1}(t) + \mu_j m_{j-1}(t) \quad (3.38)$$

This is a standard system of linear differential equations with constant coefficients that may be solved by standard methods.

If $r(j)$ is constant, and the repairman model assumptions are fulfilled, then it has been shown by Latouche that

$$E[R \mid W(R) = t] = Ct, \qquad (3.39)$$

i.e. is linear in t, with C depending upon λ and μ. See the article by Mitra (1981) for more detail concerning this problem.

3.3 REPAIRMAN MODEL IN RANDOM ENVIRONMENT

Suppose we have m machines (this may mean electric power generators, or even remote computer terminals) that, when in use, fail independently (computer terminals: apply for processing time, or data) at rate $\lambda(t)$, and, if failed, are repaired at rate $\mu(t)$; all processes are Markovian, *given* $\lambda(t)$ and $\mu(t), t \geq 0$. Now let $\lambda(t)$ and $\mu(t)$ themselves be realizations of a finite state Markov process that develops independently of the number $N(t)$ of machines down for repair. If $N(t)$ is the number down at t, and $J(t)$ is the underlying environmental state, then $\{N(t), J(t)\}$ is a bivariate Markov process, and $N(t)$ change is governed by the current level of the environment $J(t)$. The latter environment may refer to physical conditions such as heat, seismic shock, or to variations in repair effectiveness. In the case of computer terminals or communication nodes the environmental variations may be the result of changes in message transmissions or data demands under occasional crisis conditions.

The paper by Gaver *et al.* (1981) presents a systematic mathematical analysis of the general birth-and-death process in random environments, including the above repairman model as a special case. Numerical illustrations are provided. Here we present a truncated version of the solution to the first-passage time problems, utilizing a recursive or "clawing-up" mode of thinking analogous to developments earlier in this account. Restrict the discussion to just two environmental levels, denoted by $j = 1, 2$. Let the transition rate from environmental state $j = 2 \rightarrow 1$ be α, and from $j = 1 \rightarrow 2$ be β. Let $\lambda_n(j)$ be the transition rate from $(n, j) \rightarrow (n + 1, j)$, and $\mu_n(j)$ be that from $(n, j) \rightarrow (n - 1, j)$.

Let, as before, U_n be the first-passage time from n to $n + 1$, and put

$$G_n(dx; i, j) = P\{U_n \varepsilon(dx), J(U_n) = j \mid N(0) = n, J(0) = i\} \quad (3.40)$$

with $i, j = 1, 2$; the Laplace–Stieltjes transform of this measure is called

$$G_n(i, j; s) = E[e^{-sU_n}, J(U_n) = j \mid N(0) = n, J(0) = i]; \quad (3.41)$$

this is the transform of the time to pass for the first time from a state in which n are down, the environment being in state i at some initial instant ($t = 0$). to $n + 1$ down, the environment being in state j. Simple considerations of cases that may arise during the first transition give, first for $i = 1$,

$$G_n(1, j; s) = \frac{\lambda_n(1)}{d_n(1)} l_j(1) + \frac{\mu_n(1)}{d_n(1)} \sum_{k=1}^{2} G_{n-1}(1, k; s) G_n(k, j; s)$$

$$+ \frac{\alpha}{d_n(1)} G_n(2, j; s) \quad (3.42)$$

where $j = 1, 2$, the indicator function

$$l_i(l) = \begin{cases} 1 & \text{if } i = j \\ 0 & \text{otherwise} \end{cases} \quad (3.43)$$

and the denominator

$$d_n(i) = \lambda_n(1) + \mu_n(1) + \beta + s \quad (3.44)$$

Likewise, for $i = 2$,

$$G_n(2, j; s) = \frac{\lambda_n(2)}{d_n(2)} l_j(2) + \frac{\mu_n(2)}{d_n(2)} \sum_{k=1}^{2} G_{n-1}(2, k; s) G_n(k, j; s)$$

$$+ \frac{\beta}{d_n(2)} G_n(1, j; s) \quad (3.45)$$

(the equivalent of these equations in Chu and Gaver (1977) accidentally incorrectly omits the final term on the right-hand side). The above equations can in principle be solved recursively, beginning with

$$G_0(1, j; s) = \frac{\lambda_0(1)}{d_0(1)} l_j(1) + \frac{\alpha}{d_0(1)} G_0(2, j; s)$$

and (3.46)

$$G_0(2, j; s) = \frac{\lambda_0(2)}{d_0(2)} l_j(2) + \frac{\beta}{d_0(2)} G_0(1, j; s) \quad j = 1, 2$$

where

$$d_0(1) = \lambda_0(1) + \beta + s$$

and

$$d_0(2) = \lambda_0(2) + \alpha + s$$

The first-passage time T_n from $N(0) = 0$ and $J(0) = i$ ($i = 1, 2$) to $n + 1$ has the transform

$$P_n(i, j; s) = E[e^{-sT_n}, J(T_n) = j \mid N(0) = 0, J(0) = i];$$

in matrix notation,

$$\mathbf{P}_n = \mathbf{G}_0\mathbf{G}_1 \ldots \mathbf{G}_n$$

and the Laplace–Stieltjes transform of the first-passage time to $n + 1$ is $\mathbf{P}_n l$, where $l = (1, 1)^T$, a column vector.

Differentiation of expressions (3.2) and (3.5) produces recursive expressions for means, variances and higher moments. Programs have been written to evaluate these expressions numerically.

Acknowledgement

Research reported here was partially supported by the Statistical Probability Branch Office of Naval Research.

REFERENCES

AITCHISON, J., and BROWN, J. A. C. (1957). "The Log-Normal Distribution", Cambridge University Press, England.
BENDER, E. A. (1978). "An Introduction to Mathematical Modeling", John Wiley & Sons Inc., New York.
BURMAN, D. Y. (1979). An Analytic Approach to Diffusion Approximations in Queueing. Ph.D. Dissertation, NYU Courant Inst.
CHU, B. B. and GAVER, D. P. (1977). Stochastic models for repairable redundant systems susceptible to common mode failure. S. Proc. of Int. Conf. on Nuclear Systems, Reliab. Eng. and Risk Assessment, Gatlinburg, T, pp. 342–367.
CHUNG, K. L. (1967). "Markov Chains with Stationary Transition Probabilities", 2nd Ed, Springer-Verlag, New York.
COX, D. R. (1962). "Renewal Theory", Methuen Monograph, John Wiley & Sons Inc., New York.
COX, D. R. (1969). "Analysis of Binary Data", Chapman and Hall, London.
COX, D. R. and SMITH, W. L. (1961). "Queues", Methuen Monograph, John Wiley & Sons Inc., New York.

48   *Donald P. Gaver*

CRAMER, H. (1946). "Mathematical Methods of Statistics", Princeton University Press, Princeton, NJ.

DANIELS, H. E. (1954). Saddlepoint approximations in statistics. *Ann. of Math. Statistics*, **25**, pp. 631–650.

FELLER, W. (1951; 1966). "An Introduction to Probability Theory and Its Applications, Vols. I and II" John Wiley & Sons Inc., New York.

GAVER, D. P. (1962). A waiting line with interrupted service, including priorities. *J. Roy. Stat. Soc., B*, **25**, pp. 73–90.

GAVER, D. P. (1969). Highway delays resulting from flow-stopping incidents. *J. App. Prob.*, **6**, pp. 137–153.

GAVER, D. P. and JACOBS, P. A. (1981). On combinations of random loads. *S. J. Appl. Math.*, **40**, pp. 454–466.

GAVER, D. P. and LEHOCZKY, J. P. (1981). Diffusion approximations for the cooperative service of voice and data messages. *J. Appl. Prob.*, **18**, pp. 660–671.

GAVER, D. P. and SHEDLER, G. (1973). Approximate models for processor utilization in multiprogrammed computer systems. *S. J. Comput.*, **2**, No. 3, pp. 183–192.

GAVER, D. P. and THOMPSON, G. L. (1973). "Programming and Probability Models in Operations Research", Brooks-Cole Publ. Co., Monterey, CA.

GAVER, D. P., JACOBS, P. A. and LATOUCHE, G. (1981). Finite birth-and-death models in randomly changing environments. Report Interne No. 121, Lab. D'Informat. Theor., Univ. Libre de Brux.

HARRISON, J. M. and REIMAN, M. I. (1981), Reflected Brownian motion on an orthant. *Ann. Prob.*, **9**, pp. 302–308.

JACOBS, P. A. (1978). A cyclic queueing network with dependent exponential service times. *J. Appl. Prob.*, **15**, pp. 573–589.

JACOBS, P. A. (1980). Heavy traffic results for single-server queues with dependent (EARMA) service and arrival times. *Adv. Appl. Prob.*, **12**, No. 2, pp. 517–529.

KARLIN, S. and TAYLOR, H. M. (1975). "A First Course in Stochastic Processes', Academic Press, London and New York.

KLEINROCK, L. (1976). "Queueing Systems", Vols. I and II, John Wiley & Sons Inc., New York.

KLINE, M. B. and ALMOG, R. (1980). Suitability of the lognormal distribution for repair times. Proc. Second Int. Conf. on Reliability and Maintainability, Perros-Guirec-Trégastel, France, Sept. 1980.

KOBAYASHI, H. (1983), This volume.

LEVY, D. and KAHN, E. (1981). Accuracy of the Edgeworth expansion of Loss-of-Load Probability calculations in small power systems. Summer Mtg., IEEE Power Eng. Soc. (*IEEE Trans. on Power Apparatus and Systems*, to appear.)

LEWIS, P. A. W. and SHEDLER, G. S. (1979). Simulation of non-homogeneous Poisson processes by thinning. *Naval Res. Log. Quart.*, **26**, No. 3, pp. 403–413.

LUGANNANI, R. and RICE, S. (1980). Saddle point approximation for the distribution of the sum of independent random variables. *Adv. Applied Prob.*, **12**, No. 2, pp. 475–590.

McNEIL, D. R. (1970). Integral functionals of birth and death processes and related limiting distributions. *Ann. Statistics*, **41**, No. 2, pp. 480–485.

McNEIL, D. R. and SCHACH, S. (1973). Central limit analogues for Markov population processes (with discussion). *J. Roy. Stat. Soc.*, **35**, pp. 1–23.

MITRA, D. (1981). Waiting time distributions from closed queueing network models of shared-processor systems. Proc. Eighth Int. Symp. Computer Performance Modelling Measurement and Evaluation, Amsterdam.

NEUTS, M. F. (1981). "Matrix-Geometric Solutions in Stochastic Models—An Algorithmic Approach", The Johns Hopkins University Press, Baltimore, MD.

NEUTS, M. F. and MEIER, K. S. (1981). On the use of phase type distributions in reliability modelling of systems with a small number of components. *OR Spektrum*, **2**, 227–234.

NEWELL, G. F. (1971). "Applications of Queueing Theory", Chapman and Hall, London.

NEWELL, G. F. (1979). "Approximate Behavior of Tandens Queues", Springer-Verlag, Berlin.

PREGIBON, D. (1981). Logistic regression diagnostics. *Ann. Statistics*, **9**, pp. 705–724.

REIMAN, M. I. (1982). Open queueing networks in heavy traffic, *Math. of Opns. Res*, to appear.

TAKACS, L. (1962). "Introduction to the Theory of Queues", Oxford University Press, New York.

PART II

Stochastic Modeling: Queueing Models

Hisashi Kobayashi

4. Discrete-time queueing systems

4.1 INTRODUCTION

Most of the literature in queueing theory deal with continuous-time models; namely, interarrival times and service times are nonnegative real-valued variables. But there are many systems that operate on a discrete-time basis. Examples are the machine cycle of a processor, synchronous communication channels (e.g. slotted ALOHA multi-access channel), etc. In such systems, all events are allowed to occur only at reguarly spaced points in time. Aside from such discrete structures of intrinsic nature, we often find it computationally convenient to deal with discrete-time systems when we must obtain a numerical solution of a given problem. In this chapter we shall examine some fundamental issues of discrete-time point processes and related queueing systems.

4.2 POISSON SEQUENCE AND BERNOULLI SEQUENCE

The Poisson process, which plays a major role in queueing theory, has two discrete-time analogs: one is the Poisson sequence and the other is the Bernoulli sequence.

Definition 4.1: a discrete-time integer-valued process $\{X_k, k = 0, 1, 2, \ldots\}$ is called a *Poisson sequence* with parameter λ if the X_ks are IID with Poisson distribution of mean λ.

$$Pr[X_k = n] = \frac{\lambda^n}{n!} e^{-\lambda} \qquad n = 0, 1, 2, 3, \ldots$$

Definition 4.2: a discrete-time "0–1" valued process $\{X_k, k = 1, 2, 3, \ldots\}$ is called a *Bernoulli sequence* with parameter λ, if the X_ks are IID with distribution

$$Pr[X_k = 1] = \lambda, \qquad Pr[X_k = 0] = 1 - \lambda \quad \text{for all } k = 0, 1, 2, \ldots$$

In a given Bernoulli sequence $\{X_k\}$, let T_i be the distance between the $(i-1)$th and the ith variables and be equal to one. Then $\{T_i\}$ are IID with the following geometric distribution

$$Pr[T_i = n] = \lambda(1 - \lambda)^{n-1} \qquad n = 1, 2, 3, \ldots$$

A sequence of messages that is generated from an interactive terminal may be adequately modeled as a Bernoulli sequence. A Poisson sequence, on the other hand, may suitably characterize the stream of messages coming from a large number of sources. As we shall show in the examples of later sections, the Poisson sequence model will be found useful to the analysis of statistical time-division multiplexing, the slotted ALOHA random access scheme, etc.

Consider a continuous-time process $X(t)$, and let X_k be the number of points (i.e. arrivals) in the interval $(k-1)\Delta \leq t < k\Delta$. If $X(t)$ is a Poisson process with rate μ, then X_k is a Poisson sequence with mean $\lambda = \mu\Delta$. If λ is sufficiently small, so that $e^{-\lambda} \cong 1 - \lambda$, then this Poisson sequence is approximately a Bernoulli sequence as well.

An important aspect that distinguishes the Poisson and Bernoulli sequences from each other is that the former possesses the *reproductive* property, whereas the latter does not. By the reproductive property we mean that when m independent Poisson sequences with means $\lambda_i, i = 1, 2, \ldots, m$ are merged together, then the resulting sequence is also a Poisson sequence with parameter $\lambda = \sum_{i=1}^{m}\lambda_i$. However, if we merge two or more independent Bernoulli sequences, the composite stream is clearly not Bernoulli: it is not even a "0–1" sequence any more. This lack of reproductive property of the Bernoulli sequence presents a major difficulty when we attempt to construct a general theory of discrete-time queueing networks. (See Bharak-Kumar, 1980).

4.3 DISCRETE-TIME M/M/1 SYSTEM

Consider a single-server system that possesses waiting room (or buffer) of infinite capacity. We say that the system is in state n, when there are n customers at the initial epoch of a time slot. Assume that customers arrive in a given time slot according to a Bernoulli sequence of state-dependent rate $\lambda(n)$. During the time slot a customer in service will depart with probability $\mu(n)$. The scheduling discipline is assumed to be work-conserving (Kleinrock, 1976).

If we assume that in a given time slot a customer departure takes place before reception of a new customer, then $\mu(0) = 0$. Then denoting the probability that the system is in state n at the beginning of the kth slot as $P_k(n)$, we obtain the following equation

$$P_{k+1}(n) = P_k(n)[\{1 - \lambda(n)\}\{1 - \mu(n)\} + \lambda(n)\mu(n)]$$
$$+ P_k(n - 1)\lambda(n - 1)\{1 - \mu(n - 1)\}$$
$$+ P_k(n + 1)\{1 - \lambda(n + 1)\}\mu(n + 1) \qquad (4.1)$$

for $n = 1, 2, 3, \ldots$, and $k = 0, 1, 2, \ldots$ For $n = 0$ we find

$$P_{k+1}(0) = P_k(0)\{1 - \lambda(0)\} + P_k(1)\{1 - \lambda(1)\}\mu(1) \qquad (4.2)$$

We define an *equilibrium state* solution to be probability distribution $P(n)$ such that $P_k(n) = P(n)$ specifies a constant solution to Eq. (4.1). If such a distribution exists, it is unique and for each state n

$$\lim_{k \to \infty} P_k(n) = P(n) \qquad (4.3)$$

From Eq. (4.1) we readily find that $P(n)$ must satisfy the following linear difference equation

$$\mu(n + 1)\{1 - \lambda(n + 1)\}P(n + 1) - \lambda(n)\{1 - \mu(n)\}P(n)$$
$$= \mu(n)\{1 - \lambda(n)\}P(n) - \lambda(n - 1)\{1 - \mu(n - 1)\}P(n - 1)$$
$$\text{for } n = 1, 2, 3, \ldots \qquad (4.4)$$

For $n = 0$, we obtain the following balance equation from Eq. (4.2):

$$\mu(1)\{1 - \lambda(1)\}P(1) - \lambda(0)P(0) = 0 \qquad (4.5)$$

The (global) balance equation (4.4) and Eq. (4.5) immediately imply the following local (or individual) balance equations:

$$\mu(n)\{1 - \lambda(n)\}P(n) = \lambda(n - 1)\{1 - \mu(n - 1)\}P(n - 1) \qquad (4.6)$$

which is a discrete time analog of the balance equation of the continuous-time-birth-and-death process model (see for example Kobayashi, 1978a). Solving (4.6) we find then the equilibrium state distribution is given by

$$P(n) = P(0) \prod_{i=1}^{n} \frac{\lambda(i - 1)\{1 - \mu(i - 1)\}}{\mu(i)\{1 - \lambda(i)\}} \qquad (4.7)$$

provided that

$$P(0)^{-1} = 1 + \sum_{n=1}^{\infty} \prod_{i=1}^{n} \frac{\lambda(i - 1)\{1 - \mu(i - 1)\}}{\mu(i)\{1 - \lambda(i)\}} < \infty \qquad (4.8)$$

If in particular $\lambda(n)$ and $\mu(n)$ are both constants and given by λ and μ, respectively, then

$$P(0)^{-1} = \sum_{n=0}^{\infty} \left[\frac{\lambda(1 - \mu)}{\mu(1 - \lambda)} \right]^n = \frac{\mu(1 - \lambda)}{\mu - \lambda} \tag{4.9}$$

Hence the probability that the system is empty is given by

$$P(0) = \frac{\mu - \lambda}{\mu(1 - \lambda)} \tag{4.10}$$

Note that when an arrival and a departure occur in the same interval, then the system state defined above does not change. This means that the arrival and departure processes are not completely characterizable in terms of the states defined above. Thus, we introduce the following microstate (Hsu and Burke, 1976), which we denote by $(n(k), \alpha, \beta,)$ in which $n(k)$ is the number of customers in the system at time k, and α and β represent the number of arrivals and departures in the interval $(k - 1, k)$. We denote the probability of transition from $(n_1(k), \alpha_1, \beta_1)$ to $(n_2(k + 1), \alpha_2, \beta_2)$ by $P(n_1, \alpha_1, \beta_1; n_2, \alpha_2, \beta_2)$. Then one-step transition probabilities on the microstates are found to be

$$P(n, \alpha, \beta; n + 1, 1, 0) = \lambda(n)\{1 - \mu(n)\} \tag{4.11a}$$

$$P(n, \alpha, \beta; n, 0, 0) = \{1 - \lambda(n)\}\{1 - \mu(n)\} \tag{4.11b}$$

$$P(n, \alpha, \beta; n, 1, 1) = \lambda(n)\mu(n) \tag{4.11c}$$

$$P(n, \alpha, \beta; n - 1, 0, 1) = \{1 - \lambda(n)\}\mu(n) \tag{4.11d}$$

and all other one-step transition probabilities are zero.

In order to investigate the statistical properties of the departure process of the discrete-time M/M/1 system, we consider the *time-reversed* version of the original system. In the time-reversed process, the variables α and β now correspond to departures and arrivals, respectively. We define $Q(n_1, \alpha_1, \beta_1; n_2, \alpha_2, \beta_2)$ to be the one-step transition probability in reversed time, i.e. the conditional probability of finding n_2 customers at time $k - 1$, and α_2 departures and β_2 arrivals in $(k - 1, k)$ given the number of customers n_1 at time k and α_1 departures and β_1 arrivals in $(k, k + 1)$. Then, by applying Bayes' formula, it is not difficult to show (Hsu and Burke, 1976) that

$$Q(n, \alpha, \beta; n + 1, 1, 0) = \frac{P(n + 1)\mu(n + 1)\{1 - \lambda(n + 1)\}}{P(n)}$$

$$= \lambda(n)\{1 - \mu(n)\} = P(n, \alpha, \beta; n + 1, 1, 0)$$

$$\tag{4.12a}$$

Similarly we can show

$$Q(n, \alpha, \beta; n, 0, 0) = P(n, \alpha, \beta; n, 0, 0) \qquad (4.12b)$$

$$Q(n, \alpha, \beta; n, 1, 1) = P(n, \alpha, \beta; n, 1, 1) \qquad (4.12c)$$

and

$$Q(n, \alpha, \beta; n - 1, 0, 1) = P(n, \alpha, \beta; n - 1, 0, 1) \qquad (4.12d)$$

Since a homogeneous (or stationary) Markov chain is *time-reversible* if the one-step transition probabilities in reversed time are equal to those in forward time, we have proved that the chain defined on the microstates is time-reversible. Therefore, the departure process of the original system is the input of the reverse-time system, and these are statistically indistinguishable. Hence the stationary departure process of the discrete-time M/M/1 system is also a Bernoulli process with state-dependent parameter $\lambda(n)$.

Now let us consider a series of M queueing systems in tandem. Assume that the arrival process to the first queueing system is a Bernoulli sequence with constant rate λ. Then from the result obtained above we see that the departure process from the first system is also a Bernoulli sequence of rate λ. This departure process is the input process to the second queue. This argument can be repeated to show that the input to each queue in the cascaded system is Bernoulli with the same rate λ.

From the reversibility of the process we deduce an additional property of the discrete-time M/M/1 system when the arrival rate is queue-independent. Since the arrival process after time k is independent of $N(k)$, the number of customers found in the system at time k, it follows that the output process before k is independent of $N(k)$. This *independence* property implies the following important property of the cascaded M/M/1 systems: the number of customers $N_{m+1}(k)$, $N_{m+2}(k)$, ... $N_M(k)$ found in the $(m + 1)$th, $(m + 2)$th, ... Mth queues at time k depend only on the output from the mth queue prior to k, and are independent of $N_m(k)$. This independence property leads to the following decomposition of the joint state distribution:

$$Pr[N_1(k) = n_1, N_2(k) = n_2, \ldots, N_M(k) = n_M]$$
$$= P_1(n_1)P_2(n_2) \ldots P_M(n_M) \qquad (4.13)$$

4.4 DISCRETE-TIME M/G/∞ SYSTEM

It is known that in the continuous-time M/G/∞ system the departure process is also a Poisson process with the same rate as that of the input process

(Mirasol, 1963). What is a discrete-time analog of this "Poisson-input, Poisson-output" property? Consider, first, a Bernoulli arrival process in a discrete-time system with infinitely many parallel servers. Because the service times of individual customers are independent, more than one server can complete the service of its customer in a given time slot, except for the special case where all customers require exactly the same amount of service, i.e. the M/D/∞ system. Thus it is clear that the output process is no longer a "0–1" valued process.

Let us now consider a Poisson sequence of parameter λ as the arrival process. Let $\{g_k\}$ denote the probabilities that the service time is k slots long, $k = 1, 2, 3, \ldots$

Theorem 4.1: assume that the system is initially empty. Then the probability that there are n customers in the discrete-time M/G/∞ system during the kth time slot is given by

$$P(n;k) = \frac{\left\{\lambda \sum_{i=0}^{k-1} G^c(i)\right\}^n}{n!} \exp\left\{-\lambda \sum_{i=0}^{k-1} G^c(i)\right\} \qquad (4.14)$$

where $G^c(i)$ is the complement of the cumulative distribution of $\{g_i\}$:

$$G^c(i) = \Pr[\text{Service time} > i \text{ slots}]$$

$$= \sum_{j=i+1}^{\infty} g_j \qquad (4.15)$$

In the limit $k \to \infty$, the distribution (4.14) leads to the following equilibrium distribution:

$$P(n) = \lim_{k\to\infty} P(n;k) = \frac{\rho^n}{n!} e^{-\rho} \qquad (4.16)$$

where

$$\rho = \lambda \sum_{i=0}^{\infty} G^c(i) = \lambda E[\text{service time}] \qquad (4.17)$$

Furthermore, the departure process is also a Poisson sequence with parameter λ.

Proof: Choose arbitrarily a customer that has arrived by time k. (This customer may have already left the system before time k.) The uniformity of the arrival process indicates that the arrival time j of this marked customer is uniformly distributed over k slots, and he will be still in the system at time k ($\geq j$) with probability $P[\text{service time} \geq k - j + 1] = G^c(k - j)$. Therefore the probability that an arbitrarily chosen customer is still in the system at time k is

$$p = \frac{1}{k} \sum_{j=1}^{k} G^c(k - j) = \frac{1}{k} \sum_{i=0}^{k-1} G^c(i) \qquad (4.18)$$

Suppose that m customers have arrived by time k. The probability that $n(\leq m)$ customers are still in the system at time k is given by

$$\binom{m}{n} p^n (1 - p)^{m-n}, \qquad 0 \leq n \leq m \qquad (4.19)$$

For a Poisson sequence with rate λ, the number n is Poisson distributed with mean λk. Thus the probability that there are n customers in the system at time k is

$$P(n; k) = \sum_{m=n}^{\infty} \frac{(\lambda k)^m}{m!} e^{-\lambda k} \binom{m}{n} p^n (1 - p)^{m-n}$$

$$= \frac{(\lambda k p)^n}{n!} e^{-\lambda k p} \qquad (4.20)$$

By substituting (4.18) into (4.20) we obtain (4.14).

With p defined by (4.18), $1 - p$ represents the probability that an arbitrarily chosen customer that has arrived by time k has departed already. Therefore, the probability that j customers have departed by time k is obtained by substituting $1 - p$ for p in (4.20):

$$q_j(k) = \frac{\{\lambda k(1 - p)\}^j}{j!} e^{-\lambda k(1-p)} \qquad (4.21)$$

which is again a Poisson distribution, but with mean $\lambda \sum_{i=0}^{k-1} G(i)$. We can show, as we did in the previous section, that the number of customers leaving the system before time k is independent of the number of customers left at time k. This independence of the past departure process from the current state of the system is important. It follows that the number of departures in different time slots are statistically independent. Thus, we find that the number of departures in slot k is Poisson distributed with mean $\lambda \sum_{i=0}^{k-1} G(i) - \lambda \sum_{i=0}^{k-2} G(i) = \lambda G(k - 1)$. In the limit $\lim_{k \to \infty} \lambda G(k - 1) = \lambda$. [Q.E.D.]

4.5 DISCRETE-TIME M/G/1 SYSTEM

The M/G/1 system provides another case in which the discrete-time interpretation of the result known for its continuous-time model is fairly straightforward. We assume that arrivals are characterized by a *Bernoulli* sequence and that the queue discipline is any nonpreemptive work-conserving queue discipline. Then the method of imbedded Markov chains developed for the continuous-time model (see, for example, Kleinrock, 1975; Kobayashi, 1978a) is applicable here also.

Let Y_i be the number of customers in the system found just after the service completion of the ith customer and let Z_i be the number of customers entering the system during the service of that customer. Then the sequences $\{Y_i\}$ and $\{Z_i\}$ are related by the recurrence equation

$$Y_{i+1} = Y_i - 1_{(Y_i > 0)} + Z_{i+1} \tag{4.22}$$

where 1_E equals 1 if the condition E holds, and 0 otherwise. The probability generating function (PGF) of the stationary distribution $\{q_n\}$ of Y_i

$$Q(z) = \sum_{n=0}^{\infty} q_n z^n = \lim_{i \to \infty} E[z^{Y_i}] \tag{4.23}$$

must satisfy the equation

$$Q(z) = \lim_{i \to \infty} E[z^{Y_{i+1}}] = \lim_{i \to \infty} E[z^{Z_i + 1}] E[z^{Y_i - 1}(Y_i > 0)] \tag{4.24}$$

If we define $B(z)$ to be the PGF of the number of arrivals during a service period of a customer, we have

$$B(z) = \sum_{n=0}^{\infty} z^n \sum_{i=n}^{\infty} g_i \binom{i}{n} \lambda^n (1 - \lambda)^{i-n}$$

$$= G_s(1 - \lambda + \lambda z) \tag{4.25}$$

where

$$G_s(z) = \sum_{i=0}^{\infty} g_i z^i \tag{4.26}$$

is the PGF of the service time distribution $\{g_i\}$. Equations (4.24) and (4.25) readily show that

$$Q(z) = B(z)[Q(0) + z^{-1}\{Q(z) - Q(0)\}] \tag{4.27}$$

or equivalently

$$Q(z) = \frac{Q(0)B(z)(z - 1)}{z - B(z)} \tag{4.28}$$

Letting $z \to 1$ in (4.28) and noting that $B(1) = 1$ we obtain

$$Q(0) = 1 - B'(1) = 1 - \lambda E[S] = 1 - \rho$$

where ρ is the server utilization. It is well known for the continuous-time

M/G/1 system that the distribution $\{q_n\}$ is the same as the distribution $\{p_n\}$ of the number of customers observed at a *randomly* chosen instant of time, and hence equals the time-average distribution as well. This important property also holds for the discrete-time M/G/1 system. One proof of this fact makes use of the notion of supplementary variables (Cox and Miller, 1965). Thus the PGF of the distribution $\{p_n\}$ is given by

$$P(z) = Q(z) = \frac{(1 - \rho)B(z)(z - 1)}{z - B(z)} \qquad (4.29)$$

Let us now extend the result to the case of *compound* or *bulk* arrivals, by allowing more than one customer to enter the system at a time. We denote this system by $M^{(A)}/G/1$ where the variable A of the superscript represents the number of customers in an arriving group. Let λ represent the group arrival rate and suppose that the size of the ith group is an RV A_i and has the distribution

$$Pr[A_i = n] = a_n \qquad (4.30)$$

with the corresponding PGF denoted by $A(z)$:

$$A(z) = \sum_{n=0}^{\infty} a_n z^n \qquad (4.31)$$

Define a set of parameters $\{\lambda_n\}$ by

$$\lambda_n = \begin{cases} 1 - \lambda, & \text{if } n = 0 \\ a_n \lambda, & \text{if } n \geq 1 \end{cases} \qquad (4.32)$$

Then λ_n may be interpreted as the arrival rate of a group of size n. We then define the generating function $A(z)$ by

$$A(z) = \sum_{n=0}^{\infty} \lambda_n z^n = 1 - \lambda + \lambda A(z) \qquad (4.33)$$

The left composition of the function G_s of (4.26) with respect to the function A, which we denote by $B^{(A)}$

$$B^{(A)}(z) = (G_s \cdot A)(z) = G_s(A(z)) = G_s(1 - \lambda + \lambda A(z)) \qquad (4.34)$$

is the PGF of the total number of customers arriving during a service period. Formula (4.29) may then be generalized to the $M^{(A)}/G/1$ system, yielding

$$P^{(A)}(z) = \frac{(1 - \rho)B^{(A)}(z)(z - 1)}{z - B^{(A)}(z)} \qquad (4.35)$$

where the server utilization is now given by

$$\rho = \lambda E[S]E[A] \qquad (4.36)$$

Computation of the queue distribution can, in principle, be obtained by inverting the generating functions in (4.29) or (4.35). In practice, a simple recurrent relation for the $\{p_n\}$ may be derived from these equations provided $B(z)$, $B^{(A)}(z)$, and $A(z)$ are rational functions of z. (See Kobayashi, 1978a, pp. 63–66.) The average queue size and average response time are easily found from (4.29) or (4.35), together with Little's formula.

If we assume further that the queue discipline is FIFO, the waiting and response time distributions can be obtained also in terms of the generating functions, as contrasted with Laplace–Stieltjes transform solutions known for the continuous-time M/G/1 system.

4.6   DISCRETE-TIME GI/G/1 SYSTEM

In the discrete-time M/G/1 discussed above, the interarrival times $\{t_n\}$ were IID with a geometric distribution. If we allow an arbitrary distribution $\{f_k\}$

$$Pr[t_n = k(\text{time units})] = f_k \qquad k = 1, 2, 3, \ldots \qquad (4.37)$$

the resulting system is a discrete-time GI/G/1 (general independent arrivals/general service time/single server). The waiting time w_n of the nth customer satisfies the following recurrence equation:

$$w_n = \max\{0, w_{n-1} + x_n\}, \qquad (4.38)$$

where $x_n = s_n - t_n$, s_n = service time of the nth cutomer, and t_n = interarrival time between the nth and the $(n + 1)$th customer. Lindley (1952) showed that the stationary waiting time distribution

$$W(k) = \lim_{n \to G} Pr[w_n \le k] \qquad (4.39)$$

was the solution of an integral equation. In general the solution involves techniques from complex function theory. Kleinrock (1976, Section 2.6) shows that a discrete version of Lindley's integral equation can be used to compute iteratively the probability distribution of the waiting time. This computation method has the advantage that it will yield the waiting time distribution with the queue in transient states (i.e. the time-dependent nonequilibrium solution). A shortcoming of this method is that the computation efforts may be too excessive to obtain the steady state solution. Under the assumption that the interarrival time distribution $\{f_k\}$ and the service time distribution $\{g_k\}$ are both of compact support (i.e. $f_k = g_k = 0$ for $k > K$), Konheim (1975) gives a method for calculating the waiting time distribution. His solution method can be viewed as a discrete version of the spectral factorization method discussed by Smith (1953) for the continuous-time GI/G/1 system.

Let

$$F(z) = \sum_{k=0}^{\infty} f_k z^k \tag{4.40}$$

and

$$G(z) = \sum_{k=0}^{\infty} g_k z^k \tag{4.41}$$

From the assumption that $\{f_k\}$ and $\{g_k\}$ have compact support, if follows that F and G are polynomials. Form the rational function

$$S(z) = \frac{1 - F(z^{-1})G(z)}{1 - z} \tag{4.42}$$

and factor $S(z)$;

$$S(z) = S^+(z)S^-(z), \tag{4.43}$$

where

$$S^+(z) = C_1 \Pi (z - z_i^+)^{r_i}, \quad 1 < |z_i^+|, \tag{4.44}$$

and

$$S^-(z) = C_2 z^{-C_3} \Pi (z - z_i^-)^{r_i}, \quad 0 < |z_i^-| < 1 \tag{4.45}$$

The constants C_1 and C_2 are chosen so that

$$S^+(1) = 1 \tag{4.46}$$

Equations (4.44) and (4.45) provide a Wiener–Hopf-like factorization of $S(z)$: $1/S^+(z)$ is analytic inside the unit disk $\{z: |z| < 1\}$ while $1/S^-(z)$ contains the irregularities of $S(z)$ in this disk. Konheim showed that the PGF of the waiting time distribution, $G_w(z)$, is given by

$$G_w(z) = \frac{1}{S^+(z)} \tag{4.47}$$

So the determination of the stationary waiting time distribution requires the polynomial factorization, which may be cumbersome except for polynomials with low degrees.

4.7 GEOMETRIC BOUNDS ON THE WAITING-TIME DISTRIBUTION OF THE DISCRETE-TIME GI/G/1 SYSTEM

In this section (see also Kobayashi, 1974) we derive a tight upper bound on

$$W_n^c(k) = Pr[w_n > k], \qquad (4.48)$$

where w_n is the waiting time of the nth customer in a busy period. For this purpose we first discuss Chebyschev's inequality and related topics.

Chebyschev's Inequality: let Y be a nonnegative random variable with finite mean $\bar{Y}$. Then, for any $\delta > 0$,

$$Pr[Y \geq \delta] \leq \frac{\bar{Y}}{\delta} \qquad (4.49)$$

Kolmogovov generalized Chebyshev's inequality to martingales and submartingales (or semimartingales):

Definition 4.3: a random variable sequence $\{Y_n\}$ is called a *martingale* if $E[\,|Y_n|\,] < \infty$ for all n and if

$$E[Y_n \,|\, Y_1, Y_2, \ldots, Y_{n-1}] = Y_{n-1} \qquad (4.50)$$

Definition 4.4: a random variable sequence $\{Y_n\}$ is called a *submartingale* (or semimartingale if $E[\,|Y_n|\,] < \infty$ for all n and if

$$E[Y_n \,|\, Y_1, Y_2, \ldots, Y_{n-1}] \geq Y_{n-1} \qquad (4.51)$$

for all n.

Kolmogovov's Inequality for Submartingales:
Let $\{Y_n\}$ be a submartingale and $Y_n \geq 0$ for all n. Then, for any $\delta > 0$.

$$Pr\{\max[Y_1, Y_2, \ldots, Y_n] \geq \delta\} \leq \frac{\bar{Y}_n}{\delta} \qquad (4.52)$$

Proof: for given δ, denote by ξ the smallest subscript $1 \leq j \leq n$ such that $Y_j \geq \delta$ and set $\xi = 0$ if no such event occurs. Then ξ is a RV with possible values $0, 1, \ldots, n$, and the expectation of Y_n can be written as

$$\bar{Y}_n = E[Y_n] = \sum_{j=0}^{n} E[Y_n | \xi = j] Pr[\xi = j]$$

$$\geq \sum_{j=1}^{n} E[Y_n | \xi = j] Pr[\xi = j] \qquad (4.53)$$

For $j \geq 1$ the event $\{\xi = j\}$ depends only on $Y_1, Y_2, \ldots, Y_j$ and therefore

$$E[Y_n | \xi = j] = E[Y_n | Y_1, Y_2, \ldots Y_j] \geq Y_j \geq \delta \qquad (4.54)$$

By substituting (4.54) into (4.53), it is evident that

$$\sum_{j=1}^{n} Pr[\xi = j] \leq \frac{\bar{Y}_n}{\delta} \qquad (4.55)$$

However,

$$\sum_{j=1}^{n} Pr[\xi = j] = Pr[Y_j \geq \delta \quad \text{for some } j = 1, 2, \ldots, n]$$
$$= Pr[\max[Y_1, Y_2, \ldots Y_n] \geq \delta] \quad \text{[Q.E.D.]} \ (4.56)$$

We now consider a GI/G/1 queue, in which customers are served in the order of their arrivals. By solving (4.38) in succession we obtain

$$w_n = \max\{0, x_n, x_n + x_{n-1}, \ldots, x_n + x_{n-1} + \ldots + x_1\} \qquad (4.57)$$

Since the random variables $\{x_i s\}$ are IID, we can redefine w_n:

$$w_n = \max\{0, x_1, x_1 + x_2, \ldots, x_1 + x_2 + \ldots + x_n\}, \qquad (4.58)$$

which is stochastically equivalent to (4.57). We now define a sequence $\{Y_i\}$ by

$$Y_0 = 1 \qquad (4.59a)$$

and

$$Y_i = z^{x_1 + x_2 + \ldots + x_i}, \qquad i \geq 1, \qquad (4.59b)$$

where z is a real-valued parameter to be determined below. If $z > 1$, then Eqs. (4.58) and (4.59) imply

$$z^{w_n} = \max\{Y_0, Y_1, \ldots, Y_n\} \qquad (4.60)$$

We define the PGF $H(z)$ of the random variables $\{x_n\}$ by

$$H(z) = E[z^{x_n}] = E[z^{S_n - 1}]E[z^{-t_n}]$$
$$= G(z)F(z^{-1}) \qquad (4.61)$$

Here we assume that z is a real parameter in an interval I_z over which $H(z)$ is bounded. The interval I_z certainly includes $z = 1$. Since all the coefficients of $H(z)$ are nonnegative, $H''(z) > 0$ for all real $z > 0$. Furthermore $H(1) = 1$ and

$$H'(1) = E[x_n] < 0 \qquad (4.62)$$

The inequality in (4.62) comes from the requirement for the queueing system under consideration to be stable. We can also assume

$$Pr[x_n > 0] = Pr[x_n \geq 1] > 0, \tag{4.63}$$

which means that sometimes the service time exceeds the interarrival time. Otherwise there would be no queueing, and hence the waiting time would be zero with probability one, as can be readily seen from (4.38). Then it is not difficult to show that $\lim_{z \to \infty} H(z) = \infty$, and therefore $H(z)$ must have exactly one real root $z_0 > 1$ that satisfies

$$H(z_0) = 1 \tag{4.64}$$

(See Geihs and Kobayashi, 1980, for a detailed discussion of the above claim).

Let z be any value in the interval I_z that satisfies

$$H(z) \geq 1 \tag{4.65}$$

Then $E[\,|Y_n|\,] = H^n(z) < \infty$ for such z, and

$$E[Y_n | Y_1, Y_2, \ldots, Y_{n-1}] = E[z^{x_1+x_2+\cdots+x_n} | z^{x_1}, z^{x_1+x_2}, \ldots, z^{x_1+x_2+\cdots+x_{n-1}}]$$

$$= E[z^{x_n}]z^{x_1+x_2+\cdots+x_{n-1}} = H(z)Y_{n-1} \geq Y_{n-1} \tag{4.66}$$

Therefore the sequence $\{Y_n\}$ is a submartingale. By applying the Kolmogovov's inequality to this submartingale we obtain

$$Pr[w_n \geq k] = Pr[z^{w_n} \geq z^k]$$

$$= Pr[\max\{Y_0, Y_1, \ldots, Y_n\} \geq z^k]$$

$$\leq \frac{E[Y_n]}{z^k} = \{H(z)\}^n z^{-k} \tag{4.67}$$

Hence for any given n and k, the tightest bound is attained by finding

$$\min_z \{H^n(z)z^{-k}\}$$

with the constraint (4.65). If it were not for this constraint, the above minimization would always be achieved by choosing z^* such that

$$\frac{z^* H'(z^*)}{H(z^*)} = \frac{k}{n} \tag{4.68}$$

If $z^* > z_0$, then we have

$$Pr[w_n \geq k] \leq H^n(z^*)z^{*-k} \tag{4.69}$$

If $z^* < z_0$, then

$$Pr[w_n \geq k] \leq z_0^{-k} \tag{4.70}$$

Note that the bound (4.69) is valid for small n and/or large k. Therefore, in the limit $n \to \infty$, the bound (4.70) prevails and we obtain an upper bound of the tail of the waiting time distribution at equilibrium state:

$$W^c(k) = \lim_{n \to G} Pr[w_n > k]$$

$$= \lim_{n \to G} Pr[w_n \geq k + 1]$$

$$\leq z_0^{-(k+1)} \qquad (4.71)$$

4.8 GEOMETRIC UPPER AND LOWER BOUNDS ON THE WAITING-TIME DISTRIBUTION OF THE DISCRETE-TIME GI/G/1 SYSTEM

In this section we will obtain a discrete-time version of the Kingman (1970) and Ross (1974) bounds. From Eq. (4.48) the complementary distribution of the waiting time w_n is given as

$$W_n^c(k) = Pr[w_n > k]$$

$$= Pr\left[\sum_{i=1}^{j} x_i > k \quad \text{for some } j = 1, 2, \ldots, n\right] \qquad (4.72)$$

In the equilibrium state ($n \to \infty$), we have

$$W^c(k) = Pr\left[\sum_{i=1}^{j} x_i > k \quad \text{for some } j = 1, 2, \ldots\right]$$

$$= Pr[Y_j > z^k \quad \text{for some } j = 1, 2, \ldots], \qquad (4.73)$$

where Y_j is defined by (1.59) with a real-valued parameter $z > 1$.

We are now in a position to introduce an integer-values version of Wald's fundamental identity originally developed for the theory of sequential analysis (Wald, 1947).

Wald's Fundamental Identity
Let us denote the sum of the first i elements of the IID RVs $\{x_n\}$ by S_i:

$$S_i = x_1 + x_2 + \ldots + x_i \qquad (4.74)$$

Let integer ξ be the "stopping time", i.e. the variable that signifies the time that the running sum S_i falls outside of some prefixed interval, say $[A, B]$, for the first time:

$$\xi = \min\{i : S_i > B \quad \text{or} \quad S_i < A\} \qquad (4.75)$$

Assume that the RVs $\{x_n\}$ satisfy the following conditions:

(a) The mean and variance of x_n both exist and the variance is strictly positive.   (4.76)

(b) $Pr[x_n > 0] > 0$ (4.77a)

and

$Pr[x_n < 0] > 0$ (4.77b)

(c) For some $r > 1$ the expectation $E[z^{x_n}] = H(z)$ (4.78)

exists for z such that $0 < z < r$. Then, the following equality holds for all z such that $H(z) \geq 1$:

$$E[z^{S_\xi}H(z)^{-\xi}] = 1 \qquad (4.79)$$

Proof: the proof is analogous to Fisz (1936), where the continuous variable cases is treated. Let n be a positive integer; then we have

$$E[z^{S_n}] = E[z^{S_n}|\xi \leq n]Pr[\xi \leq n] + E[z^{S_n}|\xi > n]Pr[\xi > n] \quad (4.80)$$

Note that the condition $\xi > n$ includes the event $\xi = \infty$, i.e. the case where the process never stops. The fact that the stopping time is ξ implies that the set of variables $x_1, x_2, \ldots, x_\xi$ are no longer independent, since they must satisfy either $S_\mu > B$ or $S_\xi < A$. If $\xi \leq n$, then $S_n - S_\xi = x_{\mu+1} + x_{\mu+2} + \ldots + x_n$ is independent from $S_\xi = x_1 + x_2 + \ldots + x_\xi$, so we can write

$$E[z^{S_n}|\xi \leq n] = E[z^{S_\xi+(S_n-S_\xi)}|\xi \leq n] = E[z^{S_\xi}H(z)^{n-\xi}|\xi \leq n] \quad (4.81)$$

Since $E[z^{S_n}] = H(z)^n$, we now have from (4.80) and (4.81)

$$H(z)^n = E[z^{S_\xi}H(z)^{n-\xi}|\xi \leq n]Pr[\xi \leq n] + E[z^{S_n}|\xi > n]Pr[\xi > n] \quad (4.82)$$

Dividing both sides by $H(z)^n$ leads to

$$1 = E[z^{S_\xi}H(z)^{-\xi}|\xi \leq n]Pr[\xi \leq n] + E[z^{S_n}|\xi > n]H(z)^{-n}Pr[\xi > n] \quad (4.83)$$

Now, obviously

$$\lim_{n\to\infty} Pr[\xi \leq n] = 1, \qquad (4.84)$$

and

$$\lim_{n\to\infty} Pr[\xi > n] = 0 \qquad (4.85)$$

Since $\xi > n$ only if $A < S_n < B$, we find that for $z \geq 1$

$$z^A < z^{S_n} < z^B \qquad (4.86)$$

Therefore $E[z^{S_n}|\xi > n]$ is finite for any n, and using the fact $H(z) \geq 1$ and (4.85) we obtain

$$\lim_{n \to \infty} E[z^{S_n}|\xi > n]H(z)^{-n}Pr[\xi > n] = 0 \qquad (4.87)$$

Furthermore we have

$$\lim_{n \to \infty} E[z^{S_\xi}H(z)^{-\xi}|\xi \leq n]Pr[\xi \leq n] = E[z^{S_\xi}H(z)^{-\xi}] \qquad (4.88)$$

By substituting (4.87) and (4.88) into (4.83) we finally obtain the identity (4.79). [Q.E.D.]

Let $z_0 > 1$ be the characteristic root as we defined in Eq. (4.68), and let us set the interval $[A, B]$ as

$$[A, B] = (-\infty, k] \qquad (4.89)$$

where the parameter k corresponds to the argument of the complementary waiting time distribution $W^c(k)$, for which we are seeking upper and lower bounds. Then by using $H(z_0) = 1$, we can simplify (4.79) as

$$1 = E[z_0^{S_\xi}] = E[z_0^{S_\xi}|S_\xi > k]Pr[S_\xi > k] \qquad (4.90)$$

Since $\xi = \min[i:S_i > k]$ from the definitions (4.75) and (4.89), we have

$$Pr[S_\xi > k] = Pr\left[\sum_{i=1}^{\xi} x_i > k \quad \text{for some } \xi = 1, 2, 3, \ldots\right]$$
$$= W^c(k) \qquad (4.91)$$

Therefore, we can rewrite (4.90) as

$$1 = z_0^k E[z_0^{S_\xi - k}|S_\xi > k]W^c(k) \qquad (4.92)$$

In order for $S_\xi > k$ to hold, there must be some integer $j \geq 0$ such that

$$S_\xi - k = x_\xi - j \qquad (4.93)$$

Since the conditional distribution of $S_\xi - k$, given that $\xi = n$, $S_{\xi-1} = k - j$, and $S_\xi > k$ is just the conditional distribution of $x_n - j$ given that $x_n > j$, it follows by conditioning on ξ and $S_{\xi-1}$ that

$$\inf_{j \geq 0} E[z_0^{x_1 - j}|x_1 > j] \leq E[z_0^{S_\xi - k}|S_\xi > k] \leq \sup_{j \geq 0} E[z_0^{x_i - j}|x_1 > j]$$

Hence from (4.92) and (4.94) we obtain the following inequality:

$$\alpha z_0^{-k} \leq W^c(k) \leq \beta z_0^{-k}, \qquad (4.95)$$

where

$$\alpha = \inf_{j > 0} D(j), \qquad (4.96)$$

$$\beta = \sup_{j \geq 0} D(j), \tag{4.97}$$

and

$$D(j) = \frac{1}{E[z_0^{x_1-j} | x_1 > j]} \tag{4.98}$$

In order to derive a somewhat weaker but more easily computable inequality, we write

$$E[z_0^{x_1-j} | x_1 > j] = E[z_0^{s_1-(t_1+j)} | s_1 > t_1 + j]$$
$$= E[z_0^{s_1-j'} | s_1 > j'] \tag{4.99}$$

The variable j' defined by $j' = j + t_1$ is restricted to the range $j' \geq t_1$. But by varying j' over the broader range $j' > 0$, we obtain

$$\alpha^* z_0^{-k} \leq \alpha z_0^{-k} \leq W^c(k) \leq \beta z_0^{-k} \leq \beta^* z_0^{-k}, \tag{4.100}$$

where

$$\alpha^* = \inf_{j \geq 0} D^*(j), \tag{4.101}$$

$$\beta^* = \sup_{j \geq 0} D^*(j), \tag{4.102}$$

and

$$D^*(j) = \frac{1}{E[z_0^{s_1-j} | s_1 > j]} \tag{4.103}$$

The function $D(j)$ defined by (4.98) can be written as

$$D(j) = \frac{1}{z_0 E[z_0^{x-j-1} | x \geq j + 1]} \tag{4.104}$$

where we have dropped the time index n of x_n for brevity of notation. From this expression we readily find an upper limit on $D(j)$:

$$D(j) \leq z_0^{-1} < 1 \tag{4.105}$$

A similar argument can be applied to $D^*(k)$ of (4.103), hence

$$D^*(j) \leq z_0^{-1} < 1 \tag{4.106}$$

Therefore we have

$$\beta \leq \beta^* \leq z_0^{-1} < 1, \tag{4.107}$$

and therefore from (4.95)

$$W^c(k) \leq z_0^{-(k+1)} \tag{4.108}$$

which agrees, interestingly enough, to the earlier result (4.71).

In order to establish a lower bound or α (or α^*) it is necessary to make some assumptions concerning the distributional form of the RVs $\{x_n\}$ (or $\{s_n\}$).

Definition 4.5*: *IFR* (increasing failure rate) and *DFR* (decreasing failure rate). Note that our definition is different from the usual definitions of IFR and DFR (e.g. Barlow and Proschan, 1967).

The distribution, $F_x(k) = Pr[X \leq k]$ of a discrete RV X is called IFR (DFR) if and only if

$$\frac{F_x^c(j + k)}{F_x^c(k)} \tag{4.109}$$

is nonincreasing (nondecreasing) in k for all $j \geq 0$, where $F_x^c(k) = 1 - F_x(k) = Pr[X > k]$.

Now, for any function $u(X)$ such that $E[u(X)] < \infty$, and for the discrete distribution $F_x(k)$ of the underlying variable X, we find the following formula:

$$E[u(X)] = \sum_{k=-\infty}^{\infty} [u(k + 1) - u(k)]F_x^c(k) \tag{4.110}$$

Then by applying the above identity to the denominator of (4.98) we have

$$E[z_0^{x-j}|x > j] = z_0 + \sum_{k=1}^{\infty} [z_0^{k+1} - z_0^k]\frac{F^c(k + j)}{F_x^c(j)} \tag{4.111}$$

Since $z_0 > 1$, the above expression is a *nonincreasing* function of j if $F_x(k) = Pr[x_m \leq k]$ is IFR, and we find

$$\alpha = \inf_{j \geq 0} D(j) = D(0) = \frac{1}{E[z_0^{x_n}|x_n > 0]} \tag{4.112}$$

and

$$\beta = \sup_{j \geq 0} D(j) = \lim_{j \to K} D(j) \tag{4.113}$$

where K is the maximum value that the RVs $\{x_n\}$ can take, i.e.

$$K = \sup\{k : F_x^c(k) > 0\} \tag{4.114}$$

We can apply the same arguments to the DFR case just by replacing the *nonincreasing* by *nondecreasing* in the sentence that followed eq. (4.111). In this case

$$\alpha = \inf_{j \geq 0} D(j) = \lim_{j \to K} D(j) \tag{4.115}$$

and

$$\beta = \sup_{j \geq 0} D(j) = D(0) = \frac{1}{E[z_0^{x_n} | x_n > 0]} \tag{4.116}$$

It is obvious that the same arguments can be applied to the computation of α^* and β^*, when the service time distribution satisfies the condition of either IFR or DFR.

Example 4.1: the discrete-time M/M/1 System. Consider the model of Section 4.2, where the arrival rate $\lambda(n)$ and the completion rate $\mu(n)$ are constant. Hence the interarrival time distribution $\{f_k\}$ and the service time distribution $\{g_k\}$ are given by

$$f_k = \lambda(1 - \lambda)^{k-1} \qquad k = 1, 2, 3, \ldots \tag{4.117}$$

and

$$g_k = \mu(1 - \mu)^{k-1} \qquad k = 1, 2, 3, \ldots \tag{4.118}$$

with the corresponding PGFs denoted by $F(z)$ and $G(z)$, respectively:

$$F(z) = \frac{\lambda z}{1 - (1 - \lambda)z} \tag{4.119}$$

and

$$G(z) = \frac{\mu z}{1 - (1 - \mu)z} \tag{4.120}$$

Then from Eq. (4.61) we find the PGF of the RVs $\{x_n\}$

$$H(z) = F(z^{-1})G(z) = \frac{\lambda\mu}{\{1 - (1 - \mu)z\}\{1 - (1 - \lambda)z^{-1}\}} \tag{4.121}$$

from which we find that

$$H'(1) = \frac{\lambda - \mu}{\lambda\mu} < 0 \tag{4.122}$$

is required for the system to be stable. The characteristic equation $H(z) = 1$ takes the form

$$(1 - \mu)z^2 + (\lambda + \mu - 2)z + (1 - \lambda) = 0 \tag{4.123}$$

from which we readily find

$$z_0 = \frac{1 - \lambda}{1 - \mu} > 1 \tag{4.124}$$

4. *Discrete-time queueing systems*

Because of the memoryless property of the geometric distribution $\{g_k\}$, we find that $D^*(j)$ of (4.103) is independent of j:

$$D^*(j) = D^*(0) = \frac{1}{E[z_0^{s_n}]} = \frac{1}{G(z_0)} \qquad (4.125)$$

Hence from (4.101), (4.102), (4.120), (4.124), and (4.125) we obtain

$$\alpha^* = \beta^* = \frac{1}{G(z_0)} = \frac{\lambda(1-\mu)}{\mu(1-\lambda)} \qquad (4.126)$$

This also implies that

$$\alpha = \beta = \frac{\lambda(1-\mu)}{\mu(1-\lambda)} \qquad (4.127)$$

Thus the upper and lower bounds coincide, and we obtain an exact expression for $W^c(k)$.

$$W^c(k) = \frac{\lambda(1-\mu)}{\mu(1-\lambda)} z_0^{-k} \frac{\lambda}{\mu} \left(\frac{1-\mu}{1-\lambda}\right)^{k+1} \qquad (4.128)$$

The probability that a newly arriving customer does not have to wait before being served is

$$W(0) = 1 - W^c(0) = \frac{\mu-\lambda}{\mu(1-\lambda)} \qquad (4.129)$$

which is the same as $P(0)$ of (4.10).

Example 4.2: GI/M/1 and D/M/1. Suppose that we drop the Bernoulli assumption in Example 4.1. As long as the service time is geometric, the arguments that led to Eq. (4.125) hold in this more general case also. Therefore we find again

$$\alpha^* = \alpha = \beta = \beta^* = \frac{1}{G(z_0)} = \frac{1-(1-\mu)z_0\}}{\mu z_0} \qquad (4.130)$$

Hence we have

$$W^c(k) = \frac{\{1-(1-\mu)z_0\}}{\mu z_0} z_0^{-k} \qquad (4.131)$$

where $z_0(>1)$ is the unique root of

$$H(z_0) = \frac{\mu z_0}{1-(1-\mu)z_0} F(z_0^{-1}) = 1 \qquad (4.132)$$

If, for instance, we assume regular arrivals in which arrival epochs are d

units apart, i.e.

$$f_k = \begin{cases} 1 & k = d \\ 0 & k \neq d \end{cases} \tag{4.133}$$

Then

$$F(z) = z^d \tag{4.134}$$

which leads to the following characteristic equation:

$$(1 - \mu)z_0^d - z_0^{d-1} + \mu = 0 \tag{4.135}$$

It can be shown that there exists a unique root z_0 such that

$$1 < z_0 < \frac{1}{1 - \mu} \tag{4.136}$$

4.9 AN APPLICATION OF DISCRETE QUEUEING MODELS: BUFFER BEHAVIOR IN STATISTICAL TIME DIVISION MULTIPLEXING

A multiplexing technique called statistical time division multiplexing (or statistical multiplexing for short) is widely used in modern computer communications. Generation of messages or packets from data terminals tend to be intermittent (or "bursty") and the statistical multiplexor concentrates a large number of such data traffic to form a more regular stream. The resulting statistical regularity allows more efficient utilization of communication links. A conceptual diagram of statistical multiplexors is shown in Fig. 4.1.

An important design consideration of such a system is to understand the relation between the capacity of the buffer and the buffer overflow probability. Many authors have reported this type of problem. (See Chu, 1970, Schwartz, 1977, Chapters 7 and 8, Kobayashi and Konheim, 1977, and Wyner, 1974). In this section we discuss two different models using the results of the preceding sections. The first model, Model A, will make use of the discrete time M/G/1 result, whereas the second model, Model B, will be based on the geometric bounds of the GI/G/1 system that we derived in the last section.

Model A

We make the following set of assumptions in the statistical multiplexing operation of Fig. 4.1.

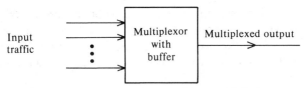

FIG. 4.1   Statistical time division multiplexing

A1: Time is divided into a sequence of intervals (or slots) of duration
     Δ (s). The amount of data that the output link can transmit in an
     interval is called the message unit: it is equal to $C \cdot \Delta$ (bits) where
     C(bits s^{-1}) is the speed of the output link

A2: The total number of message units that arrive in the kth interval is
     denoted by D_k. The RVs $\{D_k\}$ are assumed to be IID.

A3: A message unit will be sent out of the multiplexor buffer to the
     output link if and only if that message unit (and possibly some
     more) is present in the buffer at the beginning of the kth interval.

Before we start constructing a model, a few remarks are in order. Assumption A1 does not necessarily mean that the system is synchronous. Even an asynchronous system can be properly modeled by introducing an artificial slot or interval. The IID assumption of A2 is overly restrictive. For instance, if a message which starts in the kth slot is longer than Δ seconds, then this message will be counted in the slots $k, k + 1, \ldots$ This implies that the variables $D_k, D_{k+1} \ldots$ are not statistically independent, unless the message length distribution is geometric. One plausible way to cope with this difficulty will be to count all the messages that start in the kth interval to define the variable D_k. This may overestimate the variability in the buffer contents (i.e. queue size) at the multiplexor, but the statistical independence assumption will be more acceptable. Assumption A3 implies that any incoming message must wait in the buffer until the next time interval, even when the buffer is found empty upon the arrival of the message. If the interval Δ is equal to or less than the actual transfer block (i.e. character packet, or frame depending on actual implementation), then the assumption A3 is certainly realistic.

   Let us define

L_k = The amount of message units that are found outstanding in the
      multiplexor buffer at the end of the kth interval

Then assuming that the buffer capacity is sufficiently large, we find the following simple recurrence relation:

$$L_k = L_{k-1} - \mathbf{1}_{(L_{k-1}>0)} + D_k \qquad (4.138)$$

where we defined the symbol $\mathbf{1}_E$ in Eq. (4.22) of Section 4.5. Note that Eq. (4.138) is mathematically equivalent to (4.22) although our system here is not an M/G/1! Recall the subscript i in the formula (4.22) represents the sequence number of a departing customer, whereas the subscript k here is the time index.

By applying the result of (4.28) we find that the PGF of the equilibrium distribution of the buffer occupancy $\{L_k\}$ is given by

$$G_L(z) = \frac{(1 - \rho)(z - 1)G_D(z)}{z - G_D(z)} \tag{4.139}$$

provided $\rho < 1$, where ρ is the mean of D_k, and represents the output link utilization:

$$\rho = E[D_k] \tag{4.40}$$

Let us represent D (for notational convenience we now omit the time index k), as the sum of separate messages, i.e.

$$D = M_1 + M_2 + \ldots + M_A \tag{4.141}$$

where A is the number of independent messages that start arriving in a given time interval, and M_i $(1 \le i \le A)$ is the size of the ith message. Assuming that $\{M_i\}$ are IID RVs with PGF. $G_M(z)$, and the variable A has the PGF $G_A(z)$, we find

$$G_D(z) = E[z^{M_1+M_2+\ldots+M_A}]$$

$$= \sum_{n=0}^{\infty} Pr[A = n]E[z^{M_1+M_2+\ldots+M_n}|A = n]$$

$$= \sum_{n=0}^{\infty} Pr[A = n]G_M(z)^n = G_A(G_M(z)) \tag{4.142}$$

In particular

$$\rho = G_D'(1) = G_M'(1)G_A'(1) = E[M]E[A] \tag{4.143}$$

Example 4.3: suppose that there are N terminals connected to the multiplexor. Let the probability that a terminal starts generating a message in a given interval be given by $p(\ll 1)$. Clearly this Bernoulli sequence model of message generations in a given terminal is an approximation: if a terminal has started generating a message during interval k, it will be unable to start a new message during interval $k + 1$, unless the first message is shorter than the slot size Δ. This problem is exactly of the same nature as we discussed earlier in connection with the assumption A2 of Model A. Then we have

$$G_A(z) = \sum_{i=0}^{N} \binom{N}{i} p^i (1 - p)^{N-i} z^i = (pz + (1 - p))^N \qquad (4.144)$$

where we assume that the N terminals behave independently of each other. Suppose that the message length is geometrically distributed with parameter $0 \le r \le 1$:

$$Pr[M = m] = (1 - r)r^{m-1}, \qquad m = 1, 2, 3, \ldots \qquad (4.145)$$

Then the corresponding PGF is

$$G_M(z) = \frac{(1 - r)z}{1 - rz} \qquad |z| < \frac{1}{r} \qquad (4.146)$$

Hence by substituting (4.144) and (4.145) into (4.141) we obtain

$$G_D(z) = \left\{ \frac{p(1 - r)z}{1 - rz} + (1 - p) \right\}^N = \left\{ \frac{1 - p + (p - r)z}{1 - rz} \right\}^N \qquad (4.147)$$

The distribution of the buffer occupancy $\{L\}$ can then be obtained by inverting the PGF of (4.139). The recursive inversion method (see Kobayashi 1978a, Chapter 2) is appropriate for numerical evaluation of the distribution of L. We will discuss Model A further after we derive the second model:

Model B

We now replace the assumptions A1–A3 by the following corresponding ones.

B1: Time is divided, as in Model A, into a sequence of intervals of duration Δ(s). The amount of data sent out of the multiplexor in the kth interval is now denoted as R_k (message units), where the message unit is an arbitrarily chosen unit (e.g. bit, bytes, etc.)

B2: The total number of message units that arrive in the kth interval is denoted, as in Model A, by D_k. We define X_k by

$$X_k = D_k - R_k \qquad (4.148)$$

We assume that the RVs $\{X_k\}$ are IID.

B3: The multiplexor sends out data stream bit by bit and continuously as long as the buffer is not empty.

A possible usage of the assumption that the multiplexor output rate R_k can vary as time changes will be found when we deal with a communication system that will transmit data and voice. Such a communication service is often referred to as an "integrated" service. In an integrated service system the amount of output link capacity allocatable to data traffic depends on the voice traffic demands that must also be supported by the same communication link or channel. Another example is the multiplexing operation performed in a satellite ground station that accesses the satellite transponder based on the demand assigned TDMA (time division multiple access). In such a system the output bandwidth of the multiplexor varies, since a fixed amount of satellite channel capacity is dynamically partitioned among traffic bursts of variable durations to be transmitted from different earth stations. In such circumstances, the IID assumption of $\{X_k\}$ is closer to the reality than are similar assumptions for D_k or R_k. Note that we can allow dependency between D_k and R_k, insofar as $\{X_k\}$ are statistically independent.

Assumption B3 implies that any data that arrive in the interval k when the buffer is empty, will immediately be transmitted without being postponed until the $(k + 1)$th interval. This is a realistic assumption when the amount of data to be transmitted to the output link in the interval Δ is comparable or greater than the actual transport unit (bit, byte, character, or frame) to be handled by the multiplexor. Note that there is an important difference between the assumptions A3 and B3. A proper value of Δ should therefore depend on the selection of a model.

Let L_k denote, as before, the buffer occupancy at the end of the kth interval, and let us assume again that the capacity of multiplexor buffer is infinite. (The relation between the infinite capacity case and the finite capacity case will be discussed later.) Then the sequence $\{L_k\}$ satisfies the following relation:

$$L_k = \max\{0, L_{k-1} + X_k\} \qquad (4.149)$$

This equation is exactly the same as (4.38) for the GI/G/1 analysis of Section 4.6. By applying the result of (4.95), we can establish the following upper and lower bounds for the complementary distribution function of $\{L_k\}$:

$$\alpha z_0^{-n} \le F_L^c(n) \le \beta z_0^{-n} \qquad (4.150)$$

where

$$F_L^c(n) = \lim_{k \to \infty} Pr[L_k > n], \qquad (4.151)$$

$$\alpha = \inf_{m \ge 0} \frac{1}{E[z_0^{x_k - m} | x_k > m]}, \qquad (4.152)$$

$$\beta = \sup_{m \geq 0} \frac{1}{E[z_0^{x_k - m} | x_k > m]}, \qquad (4.153)$$

and $z_0 (>1)$ is the characteristic root of Eq. (4.68). If we restrict ourselves to the case where $\{D_k\}$ and $\{R_k\}$ are statistically independent, we have

$$H(z) = \lim_{k \to \infty} E[z^{x_k}] = G_D(z) G_R(z^{-1}) \qquad (4.154)$$

where

$$G_D(z) = \lim_{k \to \infty} E[z^{D_k}], \qquad (4.155)$$

and

$$G_R(z) = \lim_{k \to \infty} E[z^{R_k}] \qquad (4.156)$$

In order to compare this result with that of Model A, we consider the constant output case by setting

$$R_k = 1 \quad \text{for all } k. \qquad (4.157)$$

Then we have

$$H(z) = z^{-1} G_D(z) \qquad (4.158)$$

and the characteristic equation becomes

$$z - G_D(z) = 0 \qquad (4.159)$$

Therefore z_0 is a zero of the denominator of (4.139).

Example 4.4: geometric arrival constant output. Suppose that the number of messages arriving in an interval be geometrically distributed with parameter a, $0 < a < 1$:

$$Pr[D_k = n] = (1 - a)a^n \qquad n = 0, 1, 2, \ldots \qquad (4.160)$$

We assume that the multiplexor output rate is one message unit per interval:

$$Pr[R_k = n] = \begin{cases} 1 & n = 1 \\ 0 & n \neq 1 \end{cases} \qquad (4.161)$$

Therefore from (4.153)–(4.155) we find

$$H(z) = z^{-1} G_D(z) = \frac{1 - a}{(1 - az)z}, \qquad (4.162)$$

from which the characteristic root is readily obtained as

$$z_0 = \frac{1 - a}{a} = \rho^{-1}, \qquad (4.163)$$

where ρ is the utilization factor:

$$\rho = \frac{a}{1 - a} = E[D_k]. \tag{4.164}$$

The function $D^*(j)$ defined by (4.103) takes the following form in this example problem

$$D^*(m) = \frac{1}{E[z_0^{D_k-m}|D_k > m]} = \frac{1 - az_0}{(1 - a)z_0}, \tag{4.165}$$

which is independent of m, due to the memoryless property of the distribution of $\{D_k\}$. Therefore

$$\alpha^* = \beta^* = \frac{1 - az_0}{(1 - a)z_0} = \left(\frac{a}{1 - a}\right)^2 = z_0^{-2} \tag{4.166}$$

Therefore it follows from (4.100) that Model B gives the following result:

$$F_L^c(n) = z_0^{-2}z_0^{-n} = z_0^{-n-2} \tag{4.167}$$

Now if we apply Model A to the same problem, we have from (4.139)

$$
\begin{aligned}
G_L(z) &= \frac{(1 - \rho)(z - 1)G_D(z)}{z - G_D(z)} \\
&= \frac{(1 - z_0^{-1})(z - 1)(1 - a)/(1 - az)}{z - (1 - a)/(1 - az)} \\
&= \frac{(1 - z_0^{-1})}{(1 - zz_0^{-1})} = (1 - z_0^{-1}) \sum_{n=0}^{\infty} z_0^{-n}z^n
\end{aligned} \tag{4.168}
$$

Hence

$$Pr[L = n] = (1 - z_0^{-1})z_0^{-n}, \tag{4.169}$$

which leads to

$$F_L^c(n) = Pr[L > n] = z_0^{-n-1} \tag{4.170}$$

This is different from (4.166) by factor of z_0. This difference is attributable to the difference between the assumptions A3 and B3. Since $z_0 > 1$, the value of $F_L^c(n)$ is larger in Model A than in Model B, which agrees with our intuition. Recall that in Model A a mesage must stay in the buffer at least until the end of the interval in which the message arrives.

Example 4.5: Poisson arrivals/constant output. Suppose that the message arrivals to the multiplexor be characterized by a Poisson sequence of rate λ (message units per interval) and the multiplexor sends out one message unit per interval. Note that this model corresponds to a special case of Example 4.3 in which $\lambda = Np$ with sufficiently large N, the number of terminals, and $r = 0$, that is, all messages are of unit length. Then

$$G_D(z) = e^{\lambda(z-1)} \tag{4.171}$$

and the characteristic root is the unique real root $z_0 > 1$ that satisfies

$$e^{\lambda(z_0 - 1)} = z_0 \tag{4.172}$$

It is not difficult to show that the Poisson distribution is IFR (see Eq. (4.109)). Hence the function $D(n)$ is monotone decreasing in n, and we find

$$\alpha = D(0) = \frac{1}{E[z_0^{x_k} | x_k > 0]} \tag{4.173}$$

$$\beta = \lim_{n \to \infty} D(n) = \frac{1}{z_0} \tag{4.174}$$

Since the distribution of the RV x_k is given by

$$h_n \triangleq Pr[x_k = n] = \frac{\lambda^{n+1} e^{-\lambda}}{(n+1)!} \qquad n = -1, 0, 1, 2, \dots \tag{4.175}$$

we can evaluate the denominator of (4.173) as follows:

$$
\begin{aligned}
E[z_0^{x_k} | x_k > 0] &= \frac{\displaystyle\sum_{n=1}^{\infty} h_n z_0^n}{1 - h_{-1} - h_0} \\
&= \frac{z_0^{-1} e^{-\lambda} [e^{\lambda z_0} - 1 - \lambda z_0]}{1 - e^{-\lambda} - \lambda e^{-\lambda}}
\end{aligned} \tag{4.176}
$$

Therefore

$$\alpha = \frac{z_0(e^\lambda - \lambda - 1)}{e^{\lambda z_0} - (1 + z_0 \lambda)} = \frac{e^\lambda - \lambda - 1}{e^\lambda - \lambda - z_0^{-1}} \tag{4.177}$$

Similarly we obtain

$$\alpha^* = D^*(0) = \frac{1}{E[z_0^{D_k} | D_k > 0]} = \frac{e^\lambda - 1}{e^{\lambda z_0} - 1} = \frac{e^\lambda - 1}{z_0 e^\lambda - 1}, \tag{4.178}$$

and

$$\beta^* = \frac{1}{z_0} \tag{4.179}$$

If we apply Model A to this example case, we find

$$G_L(z) = \frac{(1 - \lambda)(z - 1)e^{\lambda(z-1)}}{z - e^{\lambda(z-1)}} = \frac{(1 - \lambda)(1 - z)}{1 - z e^{\lambda(1-z)}} \tag{4.180}$$

By applying the result of M/D/1 (Kobayashi, 1978a), we obtain the following series expansion

$$G_L(z) = (1 - \lambda)(1 - z) \sum_{j=0}^{\infty} z^j e^{j\lambda} e^{-j\lambda z} \qquad (4.181)$$

and the corresponding probability distribution is

$$Pr[L = n] = (1 - \lambda) \sum_{j=0}^{n} \frac{(-1)^{n-j}(j\lambda)^{n-j-1}(j\lambda + n - j)e^{j\lambda}}{(n - j)!}$$

$$(4.182)$$

4.10   FINITE BUFFER CAPACITY AND STATISTICAL MULTIPLEXING

In the buffer analysis of the previous section we have assumed that the buffer capacity is sufficiently large that no incoming data are lost.

We now assume that the statistical multiplexer of Model B has buffer capacity of N (message units), and let us denote the buffer occupancy by $\{L_k^*\}$. Then, instead of the recurrence relation (4.149), we find

$$L_k^* = \begin{cases} 0 & L_{k-1}^* + X_k \leq 0 \\ L_{k-1}^* + X_k & 0 \leq L_{k-1}^* + X_k \leq N \\ N & L_{k-1}^* + X_k \geq N \end{cases} \qquad (4.183)$$

or equivalently

$$L_k^* = \min[N, \max\{0, L_{k-1}^* + X_k\}] \qquad (4.184)$$

We now show that for a given sequence $\{X_k^* = (D_k - R_k)\}$ the sequences $\{L_k\}$ and $\{L_k^*\}$ satisfy the following simple inequality

$$L_k \geq L_k^* \quad \text{for all } k \qquad (4.185)$$

where L_k is, as defined by (4.148), the buffer occupancy when the capacity is assumed infinite

The relation can be proved from (4.149) and (4.184) by mathematical induction. Assuming that the systems are initially empty, i.e. $L_0 = L_0^* = 0$, we find for $k = 1$

$$L_1^* = \min[N, \max(0, X_1)] \leq \max(0, X_1) = L_1 \qquad (4.186)$$

If we assume that

$$L_k^* \leq L_k \qquad (4.187)$$

for some interval k, then for the $(k + 1)$th interval

$$
\begin{aligned}
L_{k+1}^* &= \min[N, \max\{0, L_k^* + X_{k+1}\}] \\
&\le \min[N, \max\{0, L_k + X_{k+1}\}] \qquad (4.188) \\
&\le \max\{0, L_k + X_{k+1}\} = L_{k+1}
\end{aligned}
$$

Thus the inequality (4.185) has been proved. Then it immediately follows that the complementary distribution

$$
F_{L^*}^c(n) = \lim_{k \to \infty} Pr[L^* > n] \qquad (4.189)
$$

is bounded by the corresponding expression (4.151) as follows:

$$
F_{L^*}^c(n) \le F_L^c(n) \quad \text{for all } n = 0, 1, 2, \ldots, N \qquad (4.190)
$$

Since the maximum value that $\{L_k^*\}$ can take on is N, the probability that the buffer is full in equilibrium is given by

$$
\lim_{k \to \infty} Pr[L_k^* = N] = F_{L^*}^c(N - 1) \le F_L^c(N - 1) \qquad (4.191)
$$

Then using the upper-bound (4.150) we have

$$
Pr[\text{Buffer is full}] \le \beta z_0^{-N+1} \qquad (4.192)
$$

We define the probability of overflow as the portion of lost data due to overflow:

$$
\begin{aligned}
Pr[\text{Buffer overflows}] &= \frac{\text{Offered traffic} - \text{Carried traffic}}{\text{Offered traffic}} \\
&= \frac{Q}{E[D_k]} \qquad (4.193)
\end{aligned}
$$

where Q is the average amount of data lost per interval:

$$
Q = \sum_{n=0}^{N} Pr[L_{k-1}^* = n] \left(\sum_{m=1}^{\infty} m Pr[X_k = N - n + m] \right) \qquad (4.194)
$$

Using the following (rather loose) inequalities

$$
Pr[L_{k-1}^* = n] \le F_L^c(n - 1) \le \beta z_0^{-(n-1)} \qquad n = 1, 2, 3, \ldots \qquad (4.195)
$$

and

$$
Pr[L_{k-1}^* = 0] \le 1 \qquad (4.196)
$$

we can bound Q as follows:

$$Q \leq \sum_{m=0}^{\infty} m Pr[X_K = N + m] + \beta \sum_{n=1}^{N} z_0^{-(n-1)} \sum_{m=1}^{\infty} m Pr[X_k = N - n + m]$$

$$\triangleq Q^+ \tag{4.197}$$

Therefore the buffer overflow probability is bounded as follows:

$$Pr[\text{Buffer overflows}] \leq \frac{Q^+}{E[D_k]} \tag{4.198}$$

4.11 HEAVY TRAFFIC APPROXIMATION

In the results of Sections 4.7–4.9 the characteristic root z_0 plays an important role. Numerical calculation of z_0 should present no difficulty. Newton's iteration algorithm and its variants can be directly applicable. In this section we will investigate the properties of z_0 under a special condition, i.e. when the traffic intensity of the underlying queueing system is high, and therefore z_0 is located not far from $z = 1$. We will show that an approximate value of z_0 can be explicitly given in terms of the first two moments of the interarrival times and service times.

Let us consider the following Taylor series expansion of the PGF $H(z)$ defined by (4.61):

$$H(z) = E[z^x] = H(1) + H'(1)(z - 1) + \frac{H''(1)}{2}(z - 1)^2 + O(|z - 1|^3)$$

$$= 1 + E[x](z - 1) + \{E[x^2] - E[x]\}\frac{(z - 1)^2}{2} + O(|z - 1|^3) \tag{4.199}$$

Therefore the characteristic equation

$$H(z) - 1 = (z - 1)\{E[x] + [E[x^2] - E[x]]\frac{(z - 1)}{2} + O(|z - 1|^2)\} \tag{4.200}$$

has a zero $z = 1$ and additional zero $z = z_0$, which is given by

$$E[x] + \{E[x^2] - E[x]\}\frac{(z_0 - 1)}{2} + O(|z_0 - 1|^2) = 0 \tag{4.201}$$

Note that $z_0 > 1$ since $E[x] < 0$ from (4.62). We can write

$$E[x] = E[s] - E[t] = -E[t](1 - \rho) \tag{4.202}$$

$$E[x^2] = E[x]^2 + \sigma_x^2$$
$$= E[x]^2 + \sigma_s^2 + \sigma_t^2 \tag{4.203}$$

Under the heavy traffic we have

$$E[x] \cong 0, \tag{4.204}$$

and the solution of (4.201) is approximately given by

$$z_0 \cong 1 - \frac{2E[x]}{\sigma_x^2} = 1 + \frac{2E[t](1 - \rho)}{\sigma_s^2 + \sigma_t^2} \tag{4.205}$$

In deriving this approximate formula we used the fact that $z_0 \cong 1$ under the heavy traffic condition, and hence $O(|z_0 - 1|^2)$ is negligibly small. When z_0 is close to unity, we see from (4.104) that

$$\alpha^* \cong \beta^* \cong \frac{1}{z_0} \tag{4.206}$$

Therefore, we obtain the following approximate expression for $W^c(k)$:

$$W^c(k) \cong \left[1 + \frac{2E[t](1 - \rho)}{\sigma_s^2 + \sigma_t^2} \right]^{-(k+1)}. \tag{4.207}$$

The last equation says that under heavy traffic conditions the waiting time is approximately geometrically distributed, and it depends only on the first two moments of the service time and interarrival time distributions.

Acknowledgements

A part of this lecture note was prepared while the author was with Technische Hochschule Darmstadt, Germany, being supported by the Alexander von Humboldt Foundation in the period 1979–1980. He also thanks Dr. Guy Latouche, Université Libre de Bruxelles, for his critical comments of the manuscript.

5. Diffusion approximations in queueing analysis

5.1 INTRODUCTION

In Section 4.11 we have shown that a nearly saturated queueing system (i.e. $\cong 1$) is rather insensitive to the detailed form of the arrival and service time distribution. Equation (4.207) shows, for instance, that the waiting time is geometrically distributed and its distribution parameter depends only on the first two moments. A more general theory of the heavy traffic approximation is discussed by Kingman (1962; 1965). See also Kleinrock (1976) for an expository treatment on this subject.

An approximation technique that is closely related to the heavy traffic theory is the diffusion approximation theory. The idea of approximating a discrete-state process (e.g. a random walk) by a diffusion process with continuous path is discussed by Feller (1971). Cox and Miller (1965), Gaver (1968; 1971), Newell (1971), and others discuss applications of the diffusion process approach to congestion theory. The procedure of using a diffusion process to study a queueing system—whether it be a continuous-time system or a discrete-time system—can be useful because mathematical methods associated with the continuum very often lend themselves more easily to analytical treatment than those associated with discrete coordinate axes. In the present chapter we shall present a self-contained discussion of this subject and show applications to performance analysis of some simple computer system and communication system models.

Consider a single server queueing system, and let $0 < \tau_1 < \tau_2 < \ldots$ be the arrival times of customers, and let $0 < \tau_1' < \tau_2' < \ldots$ be the departure times of these customers. Whether the time τ_i represents a continuous-time parameter or a discrete-time index is immaterial in the arguments that follow. Let $A(t)$ and $D(t)$ represent the cumulative number of arrivals and departures, respectively, up to time t:

$$A(t) = \text{Number of } \tau_i \text{ with } \tau_i \le t \tag{5.1}$$

and

$$D(t) = \text{Number of } \tau_i' \text{ with } \tau_i' \le t \tag{5.2}$$

Then the number of customers in the system at time t is given by

$$Q(t) = A(t) - D(t) \tag{5.3}$$

The change in the queue size between time t and $t + \Delta$ is

$$Q(t + \Delta) - Q(t) = \{A(T + \Delta) - A(t)\} - \{D(t + \Delta) - D(t)\} \tag{5.4}$$

which we write, for brevity of notation, as

$$\Delta Q = \Delta A - \Delta D \tag{5.5}$$

Let the interarrival times $\{t_i\}$ defined by

$$t_i = \tau_i - \tau_{i-1}, \tag{5.6}$$

and services times $\{s_i\}$ be both IID with the mean and variance given by $(\bar{t}, \sigma_t^2)$ and $(\bar{s}, \sigma_2^2)$, respectively. Note that the service times and departure times are related by the following simple equation

$$s_i = \tau_i' - \tau_{i-1}' \tag{5.7}$$

if there is no idle time between the completion of the $(i - 1)$th customer and the start of the ith customer.

Let us consider the time interval $[0, \Delta]$ by selecting the time origin $t = 0$ being equal to the beginning of the interval under consideration. Let $n(\Delta)$ represent the number of arrivals during the interval $[0, \Delta]$. Then it is clear that

$$\tau_N < \Delta \quad \text{if and only if} \quad n(\Delta) \ge N, \tag{5.8}$$

which readily implies

$$Pr[\tau_N < \Delta] = Pr[n(\Delta) \ge N] \tag{5.9}$$

Since τ_N is the sum of the IID RVs $\{t_1, t_2, \ldots, t_N\}$, the following approximate formula holds for a sufficiently large N, i.e. for a sufficiently long interval Δ.

$$Pr[\tau_N < \Delta] \cong \Phi\left(\frac{\Delta - N \cdot \bar{t}}{\sqrt{N\sigma_t^2}}\right) \tag{5.10}$$

where $\Phi(x)$ is the unit normal distribution defined by

$$\Phi(x) = \frac{1}{\sqrt{2\pi}} \int_{-\infty}^{x} \exp\left\{-\frac{u^2}{2}\right\} du \tag{5.11}$$

Then from (5.9) we find

$$Pr[n(\Delta) < N] \cong 1 - \Phi\left(\frac{\Delta - N\bar{t}}{\sqrt{N\sigma_t^2}}\right)$$

$$= \Phi\left(\frac{N - \lambda\Delta}{\sqrt{\sigma_t^2\lambda^2 N}}\right) \tag{5.12}$$

where λ is the arrival rate, i.e.

$$\lambda = 1/\bar{t} \tag{5.13}$$

For sufficiently large Δ, the distribution of $n(\Delta)$ is concentrated around $N \cong \lambda\Delta$, and we can rewrite (5.12) as

$$Pr[n(\Delta) < N] \cong \Phi\left(\frac{N - \lambda\Delta}{\sqrt{\sigma_t^2\lambda^3\Delta}}\right) \tag{5.14}$$

The last formula is directly applicable to the distribution of $\Delta A = A(t + \Delta) - \Delta A(t)$ defined by (5.5). That is, for sufficiently large Δ, the variable ΔA is distributed approximately normally with mean $\lambda\Delta$ and variance $\sigma_t^2\lambda^3\Delta$. If the traffic intensity is relatively high and the server is seldom idle, then it follows from the relation (5.7) that the variable $\Delta D = D(t + \Delta) - D(t)$ is distributed approximately normally with mean $\mu\Delta$ and variance $\sigma_s^2\mu^3\Delta$, where

$$\mu = \frac{1}{\bar{s}} \tag{5.15}$$

is the service completion rate. Furthermore, under the heavy traffic condition, the variables ΔA and ΔD can be treated as mutually independent; hence, ΔQ is approximately normally distributed with mean $\beta\Delta$ and variance $\alpha\Delta$, where

$$\beta = \lambda - \mu \tag{5.16}$$

$$\alpha = \sigma_t^2\lambda^3 + \sigma_s^2\mu^3 \tag{5.17}$$

So if we wish to approximate the discrete-valued process $Q(t)$ by a continuous process $x(t)$, its incremental change $dx(t) = x(t + dt)$ should be normally distributed with the mean βdt and variance αdt. In other words, $x(t)$ is defined by the following stochastic differential equation (Cox and Miller, 1965, p. 208)

$$dx(t) = \beta \cdot dt + z(t)\sqrt{dt} \tag{5.18}$$

where $z(t)$ is a white Gaussian process with zero mean and unit variance. If the boundary condition

$$x(t) \geq 0 \tag{5.19}$$

were not imposed, then $x(t)$ would be the Wiener–Levy process (or Brownian motion) with drift. Given the initial value x_0, the unrestricted process $x(t)$ would have the following conditional probability density function at time t:

$$f(x_0, 0; x, t) = \frac{1}{\sqrt{2\pi\alpha t}} \exp\left\{ - \frac{(x - x_0 - \beta t)^2}{2\alpha t} \right\} \qquad (5.20)$$

By taking the natural logarithm of the above equation, we write

$$\ln f = \frac{(x - x_0 - \beta t)^2}{2\alpha t} - \frac{1}{2} \ln 2\pi\alpha t \qquad (5.21)$$

Then taking the partial derivatives of $f = f(x_0, 0\,x, t)$ with respect to x and t, we have

$$\frac{\partial f}{\partial x} \frac{1}{f} = - \frac{(x - x_0 - \beta t)}{\alpha t} \qquad (5.22)$$

$$\frac{\partial^2 f}{\partial x^2} \frac{1}{f} - \left(\frac{\partial f}{\partial x}\right)^2 \frac{1}{f^2} = - \frac{1}{\alpha t} \qquad (5.23)$$

and

$$\frac{\partial f}{\partial t} \frac{1}{f} = \frac{\beta(x - x_0 - \beta t)}{\alpha t} + \frac{(x - x_0 - \beta t)^2}{2\alpha t^2} - \frac{1}{2t} \qquad (5.24)$$

We can show that the following partial differential equation should hold:

$$\frac{\partial f}{\partial t} = \frac{\alpha}{2} \frac{\partial^2 f}{\partial x^2} - \beta \frac{\partial f}{\partial x} \qquad (5.25)$$

This equation is called the Kolmogorov diffusion equation or the Fokker–Planck equation.

We now want to solve (5.25) with the constraint (5.19). A natural way to handle the nonnegativity constraint of $x(t)$ will be to treat $x = 0$ as a reflecting (or elastic) barrier. If we denote the conditional (cumulative) distribution function by $F(x_0, 0: x, t)$, the function F should satisfy the same partial differential equation as (5.25), i.e.

$$\frac{\partial F(x_0, 0; x, t)}{\partial t} = \frac{\alpha}{2} \frac{\partial^2 F(x_0, 0; x, t)}{\partial x^2} - \beta \frac{\partial F(x_0, 0; x, t)}{\partial x} \qquad (5.26)$$

and the boundary conditions are

$$\lim_{x \to \infty} F(x_0, 0; x, t) = 1 \quad \text{for all } t \qquad (5.27)$$

and

$$\lim_{x \to 0} F(x_0, 0; x, t) = 0 \quad \text{for all } t \tag{5.28}$$

The solution is obtained, for example, by the method of images (Feller 1971; Newell, 1971) as

$$F(x_0, 0; x, t) = \Phi\left(\frac{x - x_0 - \beta t}{\sqrt{\alpha t}}\right) - e^{2\beta x/\alpha}\Phi\left(-\frac{x + x_0 + \beta t}{\sqrt{\alpha t}}\right) \quad x \geq 0$$

$$\tag{5.29}$$

The numerical evaluation of this time-dependent solution is discussed in Kobayashi (1974a) in connection with the proper scaling transformations for both time t and the variable x. The equilibrium distribution is obtained readily from (5.29) as follows.

For $\lambda < \mu$ (i.e. $\beta < 0$):

$$\lim_{t \to \infty} F(x_0, 0; x, t) = 1 - e^{-2|\beta|x/\alpha} \quad x \geq 0 \tag{5.30}$$

For $\lambda > \mu$ (i.e. $\beta > 0$):

$$\lim_{t \to \infty} F(x_0, 0; x, t) = 0, \quad x \geq 0 \tag{5.31}$$

5.2 THE DIFFUSION APPROXIMATION IN A TWO-STAGE CYCLIC QUEUEING SYSTEM

Let us consider a two-stage cyclic queueing system in which the service times at Station i are IID with the mean $1/\mu$ and variance $\sigma_i^2, i = 1, 2$. Let us denote by $x(t)$ the diffusion process which approximates the queue size $n_1(t)$ (Fig. 5.1).

Then the corresponding diffusion equation that the conditional probability density function $f(x_0, 0; x, t)$ must satisfy is the same as (5.25), but now we wish to impose the condition $0 \leq x(t) \leq N$, for every $t \geq 0$, where N is the total number of customers in this closed queueing system. The parameters α and β are given, instead of (5.16) and (5.17), by

$$\beta = \mu_2 - \mu_1 \tag{5.32}$$

$$\alpha = \sigma_1^2\mu_1^3 + \sigma_2^2\mu_2^3 \tag{5.33}$$

By applying the scaling transformations

$$y = \frac{x}{|\alpha/\beta|} \quad \tau = \frac{t}{\alpha/\beta^2} \quad \text{when } \beta \neq 0 \tag{5.34}$$

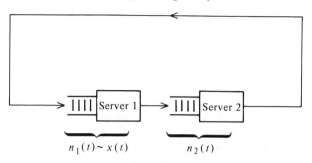

$$n_1(t) + n_2(t) = N$$

FIG. 5.1   The cyclic queueing system

and

$$y = \frac{x}{\alpha} \qquad \tau = \frac{t}{\alpha} \quad \text{when } \beta = 0 \qquad (5.35)$$

we now have the coordinate-free diffusion equation

$$\frac{\partial f}{\partial \tau}(y_0, 0; y, \tau) = \frac{1}{2}\frac{\partial^2 f}{\partial y^2}(y_0, 0; y, \tau) - \delta\frac{\partial f}{\partial y}(y_0, 0; y, \tau), \qquad (5.36)$$

where

$$\delta = \begin{cases} 1 & \text{if } \mu_2 < \mu_1 \\ 0 & \text{if } \mu_2 = \mu_1 (\text{i.e. } \beta = 0) \\ -1 & \text{if } \mu_2 > \mu_1 \end{cases} \qquad (5.37)$$

We now wish to solve (5.36) with the two reflecting barriers at $y = 0$ and at $y = b$, where

$$b = \frac{N}{|\alpha/\beta|} \quad \text{when } \beta \neq 0; \quad \text{and} \quad b = \frac{N}{\alpha} \quad \text{when } \beta = 0 \qquad (5.38)$$

Let us apply the method of separation of variables to this problem, and assume the following solution form:

$$f(y_0, 0; y, \tau) = q(y_0, y)e^{\delta y}r(\tau) \qquad (5.39)$$

and substitute this into (5.36). Then we obtain the following two equations which are interrelated via the unknown parameter d^2:

$$q''(y_0, y) + (d^2 - \delta^2)q(y_0, y) = 0 \qquad (5.40)$$

$$\dot{r}(\tau) + \frac{d^2}{2}r(\tau) = 0 \qquad (5.41)$$

where $'$ and $\cdot$ denote differentiation with respect to y and τ, respectively. Then by the standard techniques for ordinary differential equations, we find that

$$\phi(y) = Ae^{j\lambda y} + Be^{-j\lambda y} \tag{5.42}$$

satisfies (5.40), where

$$\lambda^2 = d^2 - \delta^2 \tag{5.43}$$

and $j = \sqrt{-1}$ in (5.42). By imposing the boundary condition

$$\frac{1}{2} \frac{\partial}{\partial y} f(y_0, 0; y, \tau) - \delta f(y_0, 0; y, \tau) = 0 \qquad \text{at } y = 0 \text{ and } y = b$$

$$\tag{5.44}$$

we find that the following equation must be met by λ

$$(\delta^2 + \lambda^2) \sin \lambda b = 0 \tag{5.45}$$

Furthermore the constants A and B must satisfy

$$(\delta - j\lambda)A + (\delta + j\lambda)B = 0 \tag{5.46}$$

Thus the eigenvalues that satisfy (5.45) are

$$\lambda = \pm j\delta \qquad \lambda = 0 \tag{5.47a}$$

and

$$\lambda_n = n\pi/b \qquad n = \pm 1, \pm 2, \pm 3, \ldots \tag{5.47b}$$

However, the eigenfunctions of the form (5.42) are the same for λ_n and λ_{-n}. For $\lambda = 0$ we find the corresponding function $\phi(y) = 0$. Therefore we need only consider the following eigenvalues

$$\lambda_0 = j\delta, \qquad \lambda_n = n\pi/b \qquad n = 1, 2, 3, \ldots \tag{5.48}$$

and the corresponding eigenfunctions

$$\phi_0(y) = \sqrt{\frac{2\delta}{d^{2\delta b} - 1}} \, e^{\delta y} \tag{5.49}$$

and

$$\phi_n(y) = \sqrt{\frac{2\lambda_n^2}{b(\lambda_n^2 + 1)}} \left\{ \cos \lambda_n y + \frac{\delta}{\lambda_n} \sin \lambda_n y \right\} \qquad n = 1, 2, 3, \ldots$$

$$\tag{5.50}$$

The functions $\phi_0(y)$, $\phi_1(y)$, $\phi_2(y)$, $\ldots$ are orthonormal and form complete set for all differentiable functions whose support is $[0, b]$. Once λ is determined, the corresponding solution to (5.41) is given by

$$r(\tau) = \exp\left\{-\tfrac{1}{2}(\lambda^2 + \delta^2)\tau\right\} \qquad (5.51)$$

Therefore, we arrive at the following form for a general solution

$$f(y_0, 0; y, \tau) = \sum_{n=0}^{\infty} a_n \phi_n(y) \exp\left\{\delta y - \tfrac{1}{2}(\lambda_n^2 + \delta^2)\tau\right\} \qquad (5.52)$$

The constants a_0, a_1, a_2, ... are to be determined by the initial condition $y(0) = y_0$, which is equivalent to

$$f(y_0, 0: y, \tau) = \delta(y - y_0) \qquad 0 \le y \le b \qquad (5.53)$$

where $\delta(\cdot)$ is Dirac delta function, and can be expanded in terms of eigenvalues as follows (Morse and Fishbach, 1953)

$$\delta(y - y_0) = \sum_{n=0}^{\infty} \phi_n(y)\phi_n(y_0) \qquad (5.54)$$

A careful observation of (5.52) with τ being set to zero suggests us to replace (5.53) by the following equivalent condition

$$f(y_0, 0; y, \tau) = \exp(\delta y - \delta y_0)\delta(y - y_0)$$

$$= \exp(\delta y - \delta y_0) \sum_{n=0}^{\infty} \phi_n(y)\phi_n(y_0) \qquad (5.55)$$

By comparing the coefficients of (5.52) and (5.55) we find

$$a_0 = \sqrt{\frac{2\delta}{e^{2\delta b} - 1}} \qquad (5.56)$$

and

$$a_n = \exp(-\delta y_0)\phi_n(y_0) \qquad (5.57)$$

Therefore the solution of the diffusion equation (5.36) is given by

$$f(y_0, 0; y, \tau) = \frac{2\delta e^{2\delta y}}{e^{2\delta b} - 1} + \exp\left(\delta y - \delta y_0 - \frac{\delta^2\tau}{2}\right) \sum_{n=1}^{\infty} \phi_n(y)\phi_n(y_0)\exp\left(-\frac{\lambda_n^2\tau}{2}\right) \qquad (5.58)$$

The first term of (5.58) represents the steady state probability and the second term gives the transient part. The transient term is an infinite series, but can be well approximated by finite terms, since the factor $\exp(-\lambda n^2\tau/2)$ approaches zero as n increases. Kobayashi (1974d) discusses some numerical examples, and shows how the steady state distribution is achieved as the normalized time τ elapses.

5.3   THE DIFFUSION APPROXIMATION IN A MULTIPLE ACCESS
QUEUEING SYSTEM

In the preceding sections the parameters α and β in the diffusion equation
are independent of x and t, since the arrival rate λ and completion rate μ
are both time-independent and queue-independent. In this section we will
discuss queueing systems in which the arrival rate and service completion
rate are queue-dependent. A multiple access system such as a time-sharing
system or random-access ALOHA-like system will be appropriately modeled as a queueing system with the queue-dependent arrival rate $\lambda(x)$ and
service rate $\mu(x)$. Before we proceed to a specific application we will outline the diffusion equation which is more general than (5.25) or (5.26). The
reader who is interested in a more detailed discussion is directed to Kleinrock (1976) and references therein.

We define the conditional mean and conditional variance of the time-
continuous continuous-state Markov process $x(t)$ as follows:

$$m(t\,|\,x_0, t_0) = E[x(t)\,|\,x(t_0) = x_0] \qquad t \geq t_0 \qquad (5.59)$$

and

$$v(t\,|\,x_0, t_0) = E[\{x(t) - m(t\,|\,x_0, t_0)\}^2\,|\,x(t_0) = x_0], \quad t \geq t_0 \qquad (5.60)$$

Note that $m(t_0\,|\,x_0, t_0) = x_0$ and $v(t_0\,|\,x_0, t_0) = 0$. We define the infinitesi-
mal mean $\beta(x, t)$ and infinitesimal variance $\alpha(x, t)$ as follows:

$$\beta(x, t) = \left[\frac{\partial m(\tau\,|\,x, t)}{\partial \tau}\right]_{\tau = t} \qquad (5.61)$$

$$\alpha(x, t) = \left[\frac{\partial v(\tau\,|\,x, t)}{\partial \tau}\right]_{\tau = t} \qquad (5.62)$$

These quantities can be expressed in terms of the conditional distribution
function

$$F(x_0, t_0; x_1, t_1) = P[x(t_1) \leq x_1\,|\,x(t_0) = x_0] \qquad (5.63)$$

as follows:

$$\beta(x, t) = \lim_{\Delta t \to 0} \frac{1}{\Delta t} \int_{-\infty}^{\infty} (y - x)\, dF(x, t; y; t + \Delta t) \qquad (5.64)$$

and

$$\alpha(x, t) = \lim_{\Delta t \to 0} \frac{1}{\Delta t} \int_{-\infty}^{\infty} (y - x)^2\, dF(x, t; y; t + \Delta t). \qquad (5.65)$$

Then for small Δt we have the following Taylor series expansion

$$m(t + \Delta t \,|\, x, t) = \int_{-\infty}^{\infty} y \, dF(x, t; y, t + \Delta t)$$

$$= x + \beta(x, t)\Delta t + 0(\Delta t^2) \tag{5.66}$$

and

$$v(t + \Delta t \,|\, x, t) = \int_{-\infty}^{\infty} (y - x)^2 \, dF(x, t; y, t + \Delta t)$$

$$= \alpha(x, t)\Delta t + 0(\Delta t^2) \tag{5.67}$$

Then, following the arguments given in Kleinrock (1976, pp. 69–71); we derive the following partial differential equation for f:

$$\frac{\partial f(x_0, t_0; x, t)}{\partial t} = -\frac{\partial}{\partial x}[\beta(x, t)f(x_0 t_0; x, t)] + \frac{1}{2}\frac{\partial^2}{\partial x^2}[\alpha(x, t)f(x_0, t_0; x, t)] \tag{5.68}$$

which is referred to as Kolmogorov's forward equation or the Fokker-Planck equation.

Now let us consider a multi-access system, which is schematically shown in Fig. 5.2(a). Typical characteristics of the arrival rate $\lambda(x)$ and service completion rate $\mu(x)$ are sketched in Fig. 5.2(b). If we denote the variances of interarrival times and service times by $\sigma_i^2(x)$ and $\sigma_s^2(x)$, the coefficients $\alpha(x, t)$ and $\beta(x, t)$ are

$$\beta(x, t) = \lambda(x) - \mu(x) \quad \text{for all } t \tag{5.69}$$

$$\alpha(x, t) = \sigma_i^2(x)\lambda^3(x) + \sigma_s^2(x)\mu^3(x) \quad \text{for all } t \tag{5.70}$$

Since they are time-independent, we simply write $\alpha(x)$ and $\beta(x)$ hereafter. Let x^* be such that

$$\beta(x^*) = \lambda(x^*) - \mu(x^*) = 0 \tag{5.71}$$

Linearize the function $\beta(x)$ around $x = x^*$, i.e.

$$\beta(x) = \beta(x^*) + (x - x^*)\beta'(x^*) + 0((x - x^*)^2)$$

$$= \beta_0 - \beta_1 x \tag{5.72}$$

where

$$\beta_1 = -\beta'(x^*) \tag{5.73}$$

and

$$\beta_0 = -x^*\beta'(x^*) = x^*\beta_1 \tag{5.74}$$

If we consider a narrow region around $x = x^*$ (which will be well justified as discussed below), we can approximate $\alpha(x)$ by a constant value.

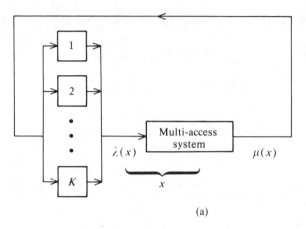

(a)

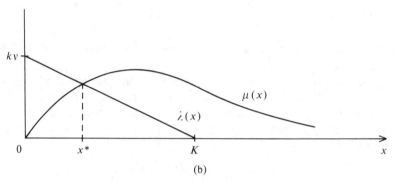

(b)

FIG. 5.2   A multi-access system

$$\alpha(x) \cong \alpha(x^*) = \lambda^3(x^*)\{\sigma_i^2(x^*) + \sigma_s^2(x^*)\}$$
$$\overset{\Delta}{=} \alpha_0. \tag{5.75}$$

Then the diffusion equation (5.68) becomes

$$\frac{\partial f}{\partial t} = \beta_1 \frac{\partial}{\partial x}\left[(x - x^*)f\right] + \frac{\alpha_0}{2}\frac{\partial^2 f}{\partial x^2} \tag{5.76}$$

The diffusion process that is characterized by the above equation is called the Ornstein–Uhlenbeck process. The time-dependent solution of the above equation is discussed in the literature (Feller, 1968; Sweet and Hardin, 1970). We denote the equilibrium-state density function by $f(x)$, i.e.

$$f(x) = \lim_{t \to \infty} f(x_0, t_0; x, t) \tag{5.77}$$

Then it is not difficult to see that

$$f(x) = \frac{1}{\sqrt{2\pi\sigma_x^2}} \exp\left\{-\frac{(x - x^*)^2}{2\sigma_x^2}\right\}$$ (5.78)

were it not for any boundary conditions, where

$$\sigma_x^2 = \frac{\alpha_0}{2\beta_1}$$ (5.79)

When we impose the reflecting barriers at $x = 0$, and $x = K$ (see Fig. 5.2(b)), then the corresponding solution is a truncated (at $x = 0$ and $x = K$) Gaussian distribution of the form (5.78).

Example 5.1: we consider the so-called machine-repairman or machine-servicing model (Kobayashi, 1978a), in which $\lambda(x)$ and $\mu(x)$ take the following form

$$\lambda(x) = (K - x)\nu \qquad 0 \le x \le K$$ (5.80)

and

$$\mu(x) = \mu \qquad 0 < x \le K$$ (5.81)

In the context of a multi-access system, ν represents the rate of request generation for terminal user. If the service times and user think times are both exponentially distributed, their variances are given by their means, $1/\mu(x)$ and $1/\lambda(x)$. Then it follows that Eq. (5.70) reduces to

$$\alpha(x) = \lambda(x) + \mu(x)$$
$$= (K - x)\nu + \mu$$ (5.82)

Since

$$\beta(x) = (K - x)\nu - \mu = (K\nu - \mu) - \nu x$$ (5.83)

we find

$$x^* = K = \frac{\mu}{\nu}$$ (5.84)

$$\beta_1 = \nu$$ (5.85)

and

$$\alpha_0 = 2\mu$$ (5.86)

Hence

$$\sigma_x^2 = \frac{\mu}{\nu}$$ (5.87)

Therefore the equilibrium distribution is given by

$$f(x) = G^{-1} \exp\left\{ - \frac{(x - K + r)^2}{2r} \right\} \tag{5.88}$$

where G is the normalization constant, and r is defined by

$$r = \frac{\mu}{\nu} \tag{5.89}$$

It will be interesting to compare this result with the exact solution that is obtained by the birth-and-death process model (Kobayashi, 1978a, p. 150). The probability, f_n, that n customers are found in service or in queue is given by the following distribution.

$$f_n = \frac{\dfrac{r^{K-n}}{(K - n)!} e^{-r}}{\displaystyle\sum_{k=0}^{K} \dfrac{r^k}{k!} e^{-r}} = \left(\sum_{k=0}^{K} \frac{r^k}{k!} \right)^{-1} \frac{r^{K-n}}{(K - n)!}, \qquad 0 \le n \le K \tag{5.90}$$

The numerator of (5.90) shows that the number of customers outside of the service system, $K - n$, has a truncated Poisson distribution. Noting that in a Poisson distribution when the mean λ is large, the distribution may be approximated by the normal distribution with mean λ and variance λ (Kobayashi, 1978a, p. 157):

$$\frac{\lambda^k}{k!} e^{-\lambda} \cong \frac{1}{\sqrt{2\pi\lambda}} \exp\left\{ -\frac{(k - \lambda)^2}{2\lambda} \right\}. \tag{5.91}$$

By substituting (5.91) into (5.90) we obtain the following approximation

$$f_n \cong \frac{\dfrac{1}{\sqrt{2\pi r}} \exp\left\{ - \dfrac{(n - K + r)^2}{2r} \right\}}{\Phi\left(\dfrac{K - r}{\sqrt{r}} \right)} \tag{5.92}$$

where $\Phi(x)$ is the unit normal distribution function defined by (5.11). The solution (5.92) is, interestingly enough, equal to (5.88) which we obtained via the diffusion approximation: the normalization constant G is approximated by

$$G \cong \sqrt{2\pi r} \; \Phi\left(\frac{K - r}{\sqrt{r}} \right) \tag{5.93}$$

6. Computational algorithms for Markovian queueing networks

6.1 INTRODUCTION

Recent progress in the area of Markovian queueing network theory has resulted in several algorithms to efficiently compute such performance measures as utilization, throughput, the average response time, and marginal distributions. The convolution method is discussed by Buzen (1973) and Reiser and Kobayashi (1975). The Polya theoretic algorithm is proposed by Kobayashi (1978a; 1976). Both algorithms deal with the normalization constant: all the performance measures are then computable via the normalization constant. Recently a new algorithm called "mean value analysis" has been discussed by Reiser and Lavenberg (1980), which is based on a simple formula for the distribution of arrival instants (Sevcik and Mitrani, 1979; Lavenberg and Reiser, 1979). Chandy and Sauer (1980) propose variations of the convolution method.

In this chapter we first review several basic formulas and then derive both the arrival time distribution and mean value analysis formulas in a much simpler way from those reported earlier by other authors. We show that the corresponding formulas for the machine-servicing model discussed by Kobayashi (1978a) are directly extendible to a general closed network. The mean value analysis is presented in terms of the normalized response times. We also show that the set of normalized response times satisfies a very simple relation.

In the final section we discuss an application of these formulas to flow control analysis of a packet-switching network, following the virtual circuit model proposed by Schwartz (1980).

99

6.2 COMPUTATIONAL ALGORITHMS AND PERFORMANCE FORMULAS

Throughout this chapter we assume a class of queueing network model, which has the equilibrium joint queue distribution of "product form". This class of queueing networks, often called Markovian networks or Jackson-type networks, is well discussed by Baskett *et al*. (1975), Kleinrock (1975), and Kobayashi (1978a; 1978b). In this chapter we limit ourselves to closed networks, since an open network with queue-dependent arrivals is representable as a weighted sum of closed networks with different population sizes (see Kobayashi, 1978a, p. 165, Eq. (3.221)).

Consider a closed queueing network with M service centers. First we discuss the case where all the servers have exponentially distributed service times with parameters μ_i, $1 \leq i \leq M$. Let K be the total number of customers in the network. Then the equilibrium-state distribution of the joint queue size $n = [n_1, n_2, \ldots, n_M]$ is given by

$$P[n] = \frac{1}{G(K, M)} \prod_{i=1}^{M} \tau_i^{n_i}, \tag{6.1}$$

where τ_i is the mean total service that is placed upon server i per unit time or per cycle. For example, if the set of parameters $\{e_i\}$ represents the average number of visits to server i per customer per cycle (where the definition of the cycle is arbitrary), then τ_i is given by

$$\tau_i = \frac{e_i}{\mu_i} \tag{6.2}$$

where $1/\mu_i$ is the mean value of the exponentially distributed sevice time at server i.

The normalization constant $G(K, M)$ of (6.1) is given by

$$G(K, M) = \sum_{n \in \mathscr{F}(K)} \prod_{i=1}^{M} \tau_i^{n_i} \tag{6.3}$$

where the sum is taken over the set of feasible states defined by

$$\mathscr{F}(K) = \left\{ n \,|\, n_i \geq 0, \quad \text{and} \quad \sum_{i=1}^{M} n_i = K \right\} \tag{6.4}$$

This normalization constant $G(K, M)$ plays a central role in numerical evaluation of the queueing model, since nearly all measures of performance can be represented in terms of the $G(K, M)$ function as shown below. The reader is directed to Kobayashi (1978a) for derivations of the following formulas.

The recursive formula:

$$G(k, m) = G(k, m - 1) + \tau_m G(k - 1, m) \quad \text{for } 1 \le k \le K, 1 \le m \le M$$

(6.5)

$$G(0, m) = 1 \quad \text{for all } m \tag{6.6}$$

$$G(k, 0) = \delta_{k,0} = \begin{cases} 1 & k = 0 \\ 0 & k > 0 \end{cases} \tag{6.7}$$

The marginal distributions:

$$\Pr[n_i = n] = \frac{G(K - n, M) - \tau_i G(K - n - 1, M)}{G(K, M)} \tau_i^n, \quad 1 \le i \le M$$

(6.8)

For $i = M$, we also have

$$\Pr[n_M = n] = \frac{G(K - n, M - 1)}{G(K, M)} \tau_M^n \tag{6.9}$$

Throughput rates:

$$\lambda_i = \theta(K, M)e_i, \quad 1 \le i \le M \tag{6.10}$$

where $\theta(K, M)$ is defined by

$$\theta(K, M) = \frac{G(K - 1, M)}{G(K, M)} \tag{6.11}$$

and we shall refer to $\theta(K, M)$ as the "normalized" throughput.

Utilization:

$$U_i = \theta(K, M)\tau_i, \quad 1 \le i \le M \tag{6.12}$$

For $i = M$, we also have

$$U_M = 1 - \frac{G(K, M - 1)}{G(K, M)} \tag{6.13}$$

Mean queue sizes:

$$N_i(K) = \frac{1}{G(K, M)} \sum_{n=1}^{K} G(K - n, M)\tau_i^n \quad 1 \le i \le M \tag{6.14}$$

The autoregressive formula:

$$G(k, M) = \frac{1}{k} \sum_{k=1}^{k} \left(\sum_{i=1}^{M} \tau_i^n \right) G(k - n, M) \qquad 1 \le k \le K \qquad (6.15)$$

The Polya formula:

$$G(K, M) = Z_{S_K}\left(\sum_{i=1}^{M} \tau_i, \sum_{i=1}^{M} \tau_i^2, \ldots, \sum_{i=1}^{M} \tau_i^K \right) \qquad (6.16)$$

where $Z_{S_K}(x_1, x_2, \ldots, x_K)$ is the cycle index polynomial of the permulation group S_K, and is given by

$$Z_{S_K}(x_1, x_2, \ldots, x_K) = \sum \frac{x_1^{i_1} x_2^{i_2} \ldots x_K^{i_K}}{i_1! 2^{i_2} i_2! \ldots K^{i_K} i_K!} \qquad (6.17)$$

where the sum is taken over the set of distinct K tuples $i_1, i_2, \ldots, i_K$ such that

$$\sum_{k=1}^{K} k i_k = K \qquad (6.18)$$

Alternatively we have the following recurrence formula

$$Z_{S_n}(x_1 x_2, \ldots, x_n) = \frac{1}{n} \sum_{k=0}^{n} x_{n-k} Z_{S_k}(x_1, x_2, \ldots, x_k) \qquad (6.19)$$

which, in turn, leads to

$$G(n, M) = \frac{1}{n} \sum_{k=0}^{n-1} x_{n-k} G(k, M) \qquad (6.20)$$

where

$$x_k = \sum_{i=1}^{M} \tau_i^k \qquad (6.21)$$

We also have

$$G(n, M) = \frac{1}{n} \sum_{k=1}^{n} x_k G(n - k, M) \qquad (6.22)$$

which is the "autoregressive formula" given in (6.15). The cycle index polynomials up to $K = 7$ are tabulated in Table 6.1.

It will be instructive to examine the formula (6.16) when all the M service stations are completely balanced, i.e. when

$$\tau_i = \tau \quad \text{for all} \quad i = 1, 2, \ldots M. \qquad (6.23)$$

TABLE 6.1   Cycle index polynomials of symmetric groups

K	Z_{S_K}
1	x_1
2	$\frac{1}{2}(x_1^2 + x_2)$
3	$\frac{1}{6}(x_1^3 + 3x_1x_2 + 2x_3)$
4	$\frac{1}{24}(x_1^4 + 6x_1^2x_2 + 3x_2^2 + 8x_1x_3 + 6x_4)$
5	$\frac{1}{120}(x_1^5 + 10x_1^3x_2 + 15x_1x_2^2 + 20x_1^2x_3 + 20x_2x_3 + 30x_1x_4 + 24x_5)$
6	$\frac{1}{720}(x_1^6 + 15x_1^4x_2 + 45x_1^2x_2^2 + 15x_2^3 + 40x_1^3x_3 + 120x_1x_2x_3 + 40x_3^2$
	$\quad + 90x_1^2x_4 + 90x_2x_4 + 144x_1x_5 + 120x_6)$
7	$\frac{1}{5040}(x_1^7 + 21x_1^5x_2 + 105x_1^3x_2^2 + 105x_1x_2^3 + 70x_1^4x_3 + 420x_1^2x_2x_3 + 210x_2^2x_3$
	$\quad + 280x_1x_3^2 + 210x_1^3x_4 + 630x_1x_2x_1 + 420x_3x_4$
	$\quad + 504x_1^2x_5 + 504x_2x_5 + 840x_1x_6 + 720x_7)$

Then we find

$$G(K, M) = Z_S(M\tau, M\tau^2, \ldots, M\tau^K)$$
$$= \tau^k Z_{S_K}(M, M, \ldots, M) \tag{6.24}$$

The last expression was obtained by factoring out τ and using the relation (6.18).

From the Polya theorem, $Z_{S_K}(M, M, \ldots, M)$ can be interpreted as the number of distinct ways of distributing K customers in the M different service stations. From an elementary combinatorial argument we see that this quantity is

$$\binom{K + M - 1}{K} = \binom{K + M - 1}{M - 1}.$$

Therefore the normalization constant becomes

$$G(K, M) = \tau^K \binom{K + M - 1}{K} \tag{6.25}$$

which of course could have been derived from (6.3).

If one or more of the servers have queue-dependent services rates, i.e. if the service completion rate of server i is given by $\mu_i(n)$ when this server holds n customers, then we define the following quantity:

$$X_i(n) = \frac{e_i^n}{\mu_i(1)\mu_i(2) \ldots \mu_i(n)} \tag{6.26}$$

and the joint queue distribution is given by

$$P[n] = \frac{1}{G(K, M)} \prod_{i=1}^{M} X_i(n_i) \quad \text{for} \quad n \in \mathcal{F}(K) \tag{6.27}$$

where

$$G(K, M) = \sum_{n \in \mathscr{F}(K)} \prod_{i=1}^{M} X_i(n_i) \qquad (6.28)$$

We then have the following formulas.

The convolution formula:

$$G(k, m) = \sum_{n=0}^{k} G(k - n, m - 1)X_m(n) \qquad (6.29)$$

with the boundary conditions (6.6) and (6.7).

The marginal distributions:
When the station M is queue-dependent,

$$Pr[n_M = n] = \frac{G(K - n, M - 1)}{G(K, M)} X_M(n). \qquad (6.30)$$

When station i is queue-dependent, we relabel staton i as station M and apply the above formula

Throughput rates:
The formulas (6.10) and (6.11) hold valid

Utilization:
The formula (6.13) holds. For an arbitrary station i, we relabel i as M, and apply the formula (6.13).

Mean queue size:

$$N_M(K) = \frac{1}{G(k, M)} \sum_{n=0}^{K} G(K - n, M - 1)X_M(n) \qquad (6.31)$$

To obtain $N_i(K)$ for arbitrary station i, we simply need relabeling.

The autoregressive formula:
If we label the servers so that the servers 1 to m ($\leq M$) are constant exponential servers,

$$G(k, m) = \frac{1}{k} \sum_{n=1}^{k} \left(\sum_{i=1}^{m} \tau_i^n \right) G(k - n, m) \qquad k = 1, 2, 3, \ldots \quad (6.32)$$

To evaluate $G(k, i)$ for $m + 1 \leq i \leq M$, the convolutional formula (6.29) is used.

The Polya formula:
If we label the serves so that the servers 1 to m ($\leq M$) are constant exponential servers,

$$G(k, m) = Z_{S_k}\left(\sum_{i=1}^{m} \tau_i, \sum_{i=1}^{m} \tau_i^2, \ldots, \sum_{i=1}^{m} \tau_i^k\right) \qquad k = 1, 2, 3, \ldots$$

(6.33)

To evaluate $G(k, i)$ for $m + 1 \leq i \leq M$, the convolutional formula (6.29) is used.

6.3 THE DISTRIBUTIONS SEEN BY ARRIVING CUSTOMERS IN A CLOSED QUEUEING NETWORK

A computational algorithm that recently has been discussed by Reiser and Lavenberg (1980) is based on the arrival theorem which states, as follows

The Arrival Theorem: in a closed Markovian network, the queue size distribution of any station observed by arriving customers is the same as the distribution that would be observed at a randomly chosen instant if that particular customer were not contributing to the system load. In other words, if we denote by $\{a_i(n; K): 0 \leq n \leq K - 1\}$ the distribution of customers found by customers arriving at station i, then

$$a_i(n; K) = p_i(n; K - 1) \qquad 0 \leq n \leq K - 1 \quad \text{for any server } i$$

(6.34)

In this section we will derive the above formula in a simple way. First, we consider the so-called machine-servicing model (or the machine-repairman model) of Fig. 6.1 which was also discussed in Chapter 5. If we denote by $p(n; K)$ the distribution of the number of customers held by the server, it is given by the following truncated Poisson distribution (see Kobayashi, 1978a, p. 150, Eq. (3.172)):

$$p(n; K) = \frac{P(K - n; r)}{Q(K; r)} \qquad 0 \leq n \leq K$$

(6.35)

where

$$P(k; \lambda) = \frac{\lambda^k}{k!} e^{-\lambda} \qquad 0 \leq k < \infty$$

(6.36)

$$Q(k; \lambda) = \sum_{i=0}^{k} P(i; \lambda) \qquad 0 \leq k < \infty$$

(6.37)

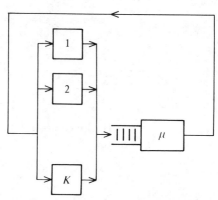

FIG. 6.1   The machine-servicing model

and r was defined by (5.89), i.e.

$$r = \frac{\mu}{\nu} \tag{6.38}$$

Over a long time interval $(t, t + \Delta)$ throughout which statistical equilibrium is to prevail, the average number of requests that arrive when the system is in state n is given by $\lambda(n)\Delta p(n; k)$, where

$$\lambda(n) = (K - n)\nu \tag{6.39}$$

As $\Delta \to \infty$, the proportion of arriving jobs that find the system in state n is therefore given by

$$a(n; K) = \frac{\lambda(n)\Delta p(n; K)}{\sum_{i=0}^{K} \lambda(i)\Delta p(i; K)}$$

$$= \frac{(K - n)p(n; K)}{\sum_{i=0}^{K-1} (K - j)p(i; K)} \qquad 0 \le n \le K - 1 \tag{6.40}$$

By substituting (6.35) into (6.40), we obtain

$$a(n; K) = \frac{(K - n)P(K - n; r)}{\sum_{j=1}^{K} jP(j; r)} \tag{6.41}$$

The denominator is equal to

$$r \sum_{j=1}^{K} \frac{r^{j-1}}{(j - 1)!} e^{-r} = rQ(K - 1; r) \tag{6.42}$$

Therefore (see Kobayashi, 1978a, p. 153, Eq. (3.181)):

$$a(n; K) = \frac{K - n}{r} \frac{P(K - n; r)}{Q(K - 1; r)} = \frac{P(K - n - 1; r)}{Q(K - 1; r)}$$

$$= p(n; K - 1) \tag{6.43}$$

Now we extend the above result to a general queueing network with product form solution. Let us consider, as before, a closed queueing network with M service centers. Let us single out the service center M; then the remaining part is a subnetwork consisting of servers 1 to M-1, as shown in Fig. 6.2.

Define the following quantities:

$$p_M(n; K) = \text{Probability that } n \text{ customers are at the center } M. \tag{6.44}$$

$$\lambda_M(n; K) = \text{The arrival rate (or throughput) at the center } M \text{ when there are } n \text{ customers at this center.} \tag{6.45}$$

The conditional throughput $\lambda_M(n; K)$ can be equated, according to Norton's theorem (Chandy *et al*, 1975; Kobayashi, 1978b, pp. 95–97), to the throughput that would be obtained through the center M, if the center M gave instantaneous services (i.e. this center is "short circuited") with the number of customers held in the network being $K - n$ (Fig. 6.3). Therefore, by use of the throughput formulas (6.10) and (6.11), we find

$$\lambda_M(n; K) = \theta(K - n; M - 1)e_M \tag{6.46}$$

where $\theta(k; M - 1)$ is the normalized throughput in the network of Fig. 6.3

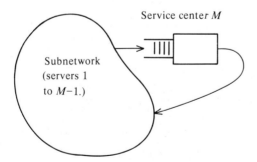

Service center M

Subnetwork
(servers 1
to M-1.)

FIG. 6.2   Decomposition of the closed network into service center M and the subnetwork

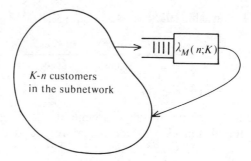

FIG. 6.3   Norton's Theorem and the conditional throughput

when the population is k:

$$\theta(k:M-1) = \frac{G(k-1, M-1)}{G(k, M-1)} \qquad (6.47)$$

Then following the arguments that led to Eq. (6.40) we find

$$a_M(n; K) = \frac{\lambda_M(n; K)p_M(n; K)}{\sum_{i=0}^{K-1} \lambda_M(i; K)p_M(i; K)} \qquad 0 \le n \le K-1 \qquad (6.48)$$

By substituting (6.46) and the marginal distribution formula (6.30) into (6.48) we find that the denominator of (6.48) is given by

$$\sum_{i=0}^{K-1} \theta(K-i; M-1)e_M \frac{G(K-i, M-1)}{G(K, M)} X_M(i) = \theta(K; M)e_M$$

$$(6.49)$$

The numerator of (6.48) becomes

$$\frac{G(K-n-1, M-1)e_M}{G(K, M)} X_M(n) \qquad (6.50)$$

Therefore we obtain

$$a_M(n; K) = \frac{G(K-n-1, M-1)}{G(K-1, M)} X_M(n) \qquad (6.51)$$

By applying the marginal distribution formula (6.30) we obtain

$$a_M(n; K) = p_M(n; K-1) \qquad (6.52)$$

Since any station can be labeled as station M, the formula (6.52) applies to any arbitrary station. That is to say, we have derived Eq. (6.34).

Now we shall derive a recursive formula for the marginal distribution $p_i(n; K)$. Consider again a long time interval $(t, t + \Delta)$ during which the system is in the equilibrium state. Then the total period during which the server i holds n customers is, on average, $\Delta p_i(n; K)$. Then the expected number of departing customers who leave behind $n - 1$ customers at the server i is given by

$$\Delta p_i(n; K)\mu_i(n) \qquad (6.53)$$

The total number of customer arrivals at the server i is the average, $\Delta\theta(K, M)e_i$, where $\theta(K, M)$ is the normalized throughput defined by (6.11). Thus, the expected total number of customers who find $n - 1$ customers upon arrival at the server i is given by

$$\Delta\theta(K, M)e_i a_i(n - 1; K) \qquad (6.54)$$

In any stable queueing system the number of arrivals that change the queue size from $n - 1$ to n should balance, in the long run, with the number of departures that drive the system from state n to state $n - 1$. Therefore, by applying the strong law of large numbers to the quantities defined by (6.53) and (6.54), we find

$$p_i(n; K)\mu_i(n) = \theta(K, M)e_i a_i(n - 1; K) \qquad (6.55)$$

Then by making use of the formula (6.34), we obtain

$$p_i(n; K) = \theta(K, M) \frac{e_i}{\mu_i(n)} p_i(n - 1; K - 1) \quad 1 \le n \le K \qquad (6.56)$$

Alternatively, we can derive this equation directly from the marginal distribution formula (6.30) as discussed by Reiser (1981).

6.4   MEAN VALUE ANALYSIS

Let us consider again the machine-servicing model of Fig. 6.1. From Eq. (6.35) we obtain

$$p(n; K) = \frac{Q(K - 1; r)}{Q(K; r)} \frac{P(K - n; r)}{Q(K - 1; r)}$$

$$= \rho(K)p(n - 1; K - 1) \qquad (6.57)$$

where

$$\rho(K) = \frac{Q(K - 1; r)}{Q(K; r)} = 1 - p(0; K) \qquad (6.58)$$

is the server utilization. Then the mean number of customers held in the system is

$$N(K) = \sum_{n=1}^{K} np(n; K) = \rho(K)\left[\sum_{n=1}^{K} np(n - 1; K - 1)\right]$$

$$= \rho(K)[N(K - 1) + 1] \qquad (6.59)$$

Let $T(K)$ denote the mean response time at the server, and define

$$Y(K) = \mu T(K) \qquad (6.60)$$

which we call the "normalized" response time (see Kobayashi, 1978a, Eqs. (3.153) and (3.175)). Since the throughput is given by $\mu\rho(K)$, we find from Little's formula

$$N(K) = \mu\rho(K)T(K) = \rho(K)Y(K) \qquad (6.61)$$

By applying Little's formula to the entire system of Fig.6.1 we have

$$K = \mu\rho(K)\left[T(K) + \frac{1}{\nu}\right]$$

$$= \rho(K)[Y(K) + r] \qquad (6.62)$$

which leads to

$$Y(K) = \frac{K}{\rho(K)} - r \qquad (6.63)$$

(see Kobayashi, 1978a, Eq. (3.175)). From (6.59) we find

$$Y(K) = \rho(K - 1)Y(K - 1) + 1 \qquad (6.64)$$

Thus Eqs (6.63) and (6.64) define a simple recursive formula for the pair $\langle Y(K), \rho(K)\rangle$. Alternatively we provide a recursion to $Y(K)$ and $\rho(K)$ separately:

$$Y(K) = K - \frac{(K - 1)r}{Y(K - 1) + r} \qquad (6.65)$$

and

$$\rho(K) = \frac{K}{K + r[1 - \rho)K - 1)]} \qquad (6.66)$$

The formula (6.65) can be derived directly from a simple recursion that holds for $p(0; K)$ (see Kobayashi, 1978a, p. 151 and 154).

We now extend the above result to a general closed queueing network. Suppose that the closed network consisting of servers 1 to M has constant service rates:

$$\frac{e_i}{\mu_i(n)} = \tau_i \quad \text{for all} \quad i = 1, 2, \ldots, M \tag{6.67}$$

Then the mean number of customers held in the server i satisfies the following simple relation

$$N_i(K) = \theta(K, M)\tau_i[N_i(K - 1) + 1], \quad 1 \leq i \leq M \tag{6.68}$$

which is a generalization of Eq. (6.59). Equation (6.60) follows from (6.34) and (6.54) as reported by Reiser and Lavenberg (1980). If we denote the mean response time at server i by $T_i(K)$,

$$N_i(K) = \theta(K, M)e_iT_i(K) = \theta(K, M)\tau_iY_i(K) \quad 1 \leq i \leq M \tag{6.69}$$

where $Y_i(K)$ is the normalized response time, i.e.

$$Y_i(K) = \mu_iT_i(K) \quad 1 \leq i \leq M \tag{6.70}$$

Summing (6.69) over i we obtain

$$K = \theta(K, M) \sum_{i=1}^{M} \tau_iY_i(K) \tag{6.71}$$

The recursive equation (6.68) can be transformed into one for $Y_i(K)$ by using (6.69):

$$Y_i(K) = \theta(K - 1, M)\tau_iY_i(K - 1) + 1 \quad 1 \leq i \leq M \tag{6.72}$$

Equations (6.71) and (6.72) together define a recursive formula for $\{\theta(K, M); Y_i(K), 1 \leq i \leq M\}$. By eliminating $\theta(K, M)$, we find

$$Y_i(K) = 1 + \frac{(K - 1)\tau_iY_i(K - 1)}{\sum_{j=1}^{M} \tau_jY_j(K - 1)} \quad 1 \leq i \leq M \tag{6.73}$$

We can also write

$$Y_i(K) = K - \frac{(K - 1) \sum_{j \neq i} \tau_iY_j(K - 1)}{\sum_{j=1}^{M} \tau_jY_j(K - 1)} \quad 1 \leq i \leq M \tag{6.74}$$

By summing (6.74) which resembles (6.65) over i, we find the following interesting relation between the set of normalized response times in the network:

$$\sum_{i=1}^{M} Y_i(K) = M + K - 1 \tag{6.75}$$

which we name the "response time identity". We will show later in Section 6.5 how to make use of this formula.

By multiplying both sides of (6.72) by τ_i^n and summing over i, we obtain

$$\sum_{i=1}^{M} \tau_i^n Y_i(K) = \theta(K - 1, M) \sum_{i=1}^{M} \tau_i^{n+1} Y_i(K - 1) + x_n \quad \text{for} \quad n = 0, 1, 2, \ldots$$

(6.76)

where x_n is defined by (6.21). For $n = 0$, the above formula reduces to (6.75) because of (6.71). It will be instructive to note that the recursive relation for the normalized throughput is given by

$$\frac{K}{\theta(K, M)} = \theta(K - 1, M) \sum_{i=1}^{M} \tau_i^2 Y_i(K - 1) + x_1 \qquad (6.77)$$

When some of the servers have queue-dependent service rates, we can no longer use the simple relation given by (6.69). Instead we must deal with the recursive formula for the marginal distribution (6.57). The following set of equations specify the recursions for the response time $T_i(K)$, the normalized throughput $\theta(K, M)$ and the set of marginal distributions:

$$T_i(K) = \sum_{n=1}^{K} \frac{n}{\mu_i(n + 1)} p_i(n - 1) \qquad 1 \le i \le M \qquad (6.78)$$

$$\theta(K, M) = \frac{K}{\sum_{i=1}^{K} e_i T_i(k)} \qquad (6.79)$$

$$p_i(n; K) = \theta(K, M) \frac{e_i}{\mu_i(n + 1)} p_i(n - 1; K - 1) \quad 1 \le i \le M \quad 1 \le n \le K$$

(6.80)

$$p_i(0; K) = 1 - \sum_{n=1}^{K} p_i(n; K) \qquad 1 \le i \le M \qquad (6.81)$$

The reader is directed to Reiser (1981) for further discussion on this algorithm.

6.5 AN APPLICATION: THE WINDOW FLOW CONTROL ANALYSIS IN PACKET-SWITCHING NETWORKS

Applications of computational algorithms to computer performance analysis are reported by various authors. The reader is directed to Kobayashi (1978a), for detailed discussions of how these formulas may be applied to performance prediction of an interactive virtual storage system. In this section we illustrate another interesting application example. Schwartz (1980a; b) discusses the performance analysis of virtual flow control in a packet-switching network. A packet (a customer) that originates at the source S traverses a series of M-links, and then arrives at the destination D. The transmission time of link i is exponentially distributed with the mean $1/\mu_i$, $i = 1, 2, \ldots, M$. In other words, the service rate μ_i represents the effective capacity allocated to this virtual circuit. If there is no control of packet flows, then the packet stream from the source S may be adequately modeled as a Poisson process of constant rate, say λ. In the so-called window flow control mechanism (see Schwartz, 1980a, and references therein), the flow rate will be adjusted so that the rate depends on n, the total number of packets in transit over this virtual circuit, as follows:

$$\lambda(n) = \begin{cases} \lambda & 0 \le n < W \\ 0 & n = W \end{cases} \tag{6.82}$$

where the integer W is called the window size. Then at any given instant $n = n_1 + n_2 + \ldots + n_M$ is at most W. Such an M-stage queueing system can be represented as the closed queueing system of Fig. 6.4, where a fictitious server, the $(M + 1)$ server, with the service rate $\mu_{M+1}(n)$, is inserted in the reverse channel (or acknowledgment channel):

$$\mu_{M+1}(n) = \lambda(W - n) \tag{6.83}$$

Since the number of visits to the $M + 1$ stations in this cyclic queueing system are equal, we can set without loss of generality that

$$e_i = 1 \quad \text{for all} \quad 1 \le i \le M + 1 \tag{6.84}$$

For the arrival rate function specified by (6.82), the corresponding closed

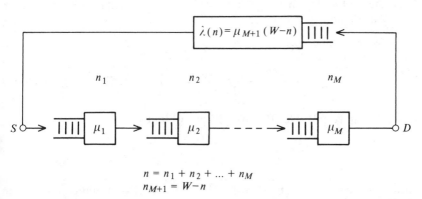

FIG. 6.4 The Queueing Model of a virtual circuit with end-to-end flow control

(M + 1) stage cyclic queueing network of Fig. 6.4 becomes a simple network with queue-independent exponential servers with rates $\mu_1, \mu_2, \ldots,$ μ_M, and μ_{M+1}, where

$$\mu_{M+1} = \lambda \tag{6.85}$$

Then any of the formulas (6.5) to (6.22) is applicable to this network of (M + 1) stations and W customers, So $G(W, M + 1)$, for instance, can be evaluated by

$$G(W, M + 1) = Z_{S_W}(x_1, x_2, \ldots x_k, \ldots, x_W) \tag{6.86}$$

where

$$x_k = \sum_{i=1}^{M+1} \mu_i^{-k} \tag{6.87}$$

If we allow the flow control rate function $\lambda(n)$ to be an arbitrary function of n, we define the function $\Lambda(n)$:

$$\Lambda(0) = 1 \tag{6.88}$$

$$\Lambda(n) = \lambda(0)\lambda(1) \ldots \lambda(n - 1) \qquad n \geq 1 \tag{6.89}$$

From the relation (6.83) and the definition (6.26) we may write

$$X_{M+1}(n) = \frac{\Lambda(W - n)}{\Lambda(W)} \qquad 0 \leq n \leq W \tag{6.90}$$

Then, from the marginal distribution formula, we find

$$\Pr[n_{M+1} = k] = \frac{G(W - k, M)}{G(W, M + 1)} X_{M+1}(k) \qquad 0 \leq k \leq W \tag{6.91}$$

Then the probability that n packets are held in the virtual circuit (i.e. from S to D in Fig. 6.4) is given by

$$p_n = \Pr[n_{M+1} = W - n]$$

$$= \frac{G(n, M)}{G(W, M + 1)} \frac{\Lambda(n)}{\Lambda(W)} \qquad 0 \leq n \leq W \tag{6.92}$$

By summing (6.92) over $0 \leq n \leq W$, we obtain

$$G(W, M + 1) = \frac{1}{\Lambda(W)} \sum_{n=0}^{W} G(n, M)\Lambda(n) \tag{6.93}$$

Note that $G(n, M)$ on the right-hand side is directly computable from the Polya formula (6.16), since the subnetwork consisting of the servers 1 to M

has all the servicing rates fixed, i.e.

$$G(n, M) = Z_{S_n}\left(\sum_{i=1}^{M} \mu_i^{-1}, \sum_{i=1}^{M} \mu_i^{-2}, \ldots \sum_{i=1}^{M} \mu_i^{-n}\right) \qquad 0 \le n \le W$$
$$(6.94)$$

It is worthwhile to remark that Eq. (6.93) corresponds to the case where we set $k = W$ in the convolution formula

$$G(k, M + 1) = \sum_{n=0}^{k} G(k - n, M)X_{M+1}(n)$$

$$= \frac{1}{\Lambda(W)} \sum_{j=0}^{k} G(j, M)\Lambda(W - k + j) \qquad (6.95)$$

Throughput, which is equivalent to $\theta(W, M + 1)$ in this case, is given from (6.11):

$$\theta(W, M + 1) = \frac{G(W - 1, M + 1)}{G(W, M + 1)} \qquad (6.96)$$

Similarly we obtain

$$E[n_{M+1}] = \sum_{n=0}^{W} G(W - n, M)nX_{M+1}(n)$$

$$= W - \frac{1}{G(W, M + 1)\Lambda(W)} \sum_{k=0}^{W} G(k, M)k\,\Lambda(k) \qquad (6.97)$$

Thus the expectation of the number of packets in the virtual circuit is

$$N(W) = \frac{\sum_{k=1}^{W} k G(k, M)\Lambda(k)}{\sum_{k=0}^{W} G(k, M)\Lambda(k)} \qquad (6.98)$$

Thus the end-to-end delay of this virtual circuit with window size W is given by

$$T(W) = \frac{N(W)}{\theta(W, M + 1)} = \frac{\sum_{k=1}^{W} k G(k, M)\Lambda(k)}{\sum_{k=0}^{W} G(k, M)\Lambda(k)} \qquad (6.99)$$

Special case:

If all the M links of the virtual circuit have an equivalent capacity, i.e. if

$$\mu_i = \mu \quad \text{for} \quad 1 \le i \le M \qquad (6.100)$$

then the formulas (6.24) and (6.25) hold, and we find

$$G(n, M + 1) = \frac{1}{\Lambda(W)} \sum_{j=0}^{n} \mu^{-j} \binom{j + M - 1}{j} \Lambda(W - n + j) \quad (6.101)$$

Similarly we find explicit expressions for the throughput $\theta(W, M + 1)$, the average number of packets on transit $N(W)$, and the end-to-end delay $T(W)$, by substituting (6.101) into (6.96), (6.98) and (6.99).

If we take out the server $M + 1$ in Figure 6.4 as we did for the server in M in Fig. 6.2, the resulting subnetwork is an M stage cyclic queue with identical exponential servers with rate μ. If we denote the throughput of this network by $\theta(n, M)$, when the population size is n, then by setting $\tau_i = 1/\mu$ in (6.69) for all i we find

$$Y_i(n) = \frac{n\mu}{M\theta(n, M)} \quad 1 \le i \le M \quad (6.102)$$

By substituting this into the response time identity

$$\sum_{i=1}^{M} Y_i(n) = M + n - 1 \quad (6.103)$$

we find

$$\theta(n, M) = \frac{n\mu}{M + n - 1} \quad (6.104)$$

Alternatively, we could have derived the last formula from (6.11) and (6.25). Note that in view of the notation defined by (6.45) and (6.46), $\theta(n, M)$ may be written as $\lambda_{M+1}(W - n; M)$:

$$\theta(n, M) = \frac{\mu^{-(n-1)}\binom{n + M - 2}{n - 1}}{\mu^{-n}\binom{n + M - 1}{n}} = \frac{n\mu}{M + n - 1} \quad (6.105)$$

Therefore, the closed queueing model of Fig. 6.4 will be reduced to that of Fig. 6.5 when the condition (6.100) holds.

If we relabel the two service stations in this equivalent cyclic queueing model by server 1^* and server 2^*, and denote

$$X_1^*(n) = \frac{1}{\lambda(W - 1)\lambda(W - 2) \ldots \lambda(W - n)} = \frac{\Lambda(W - n)}{\Lambda(W)}$$

$$(6.106)$$

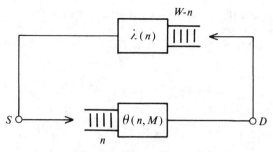

FIG. 6.5   A queueing model equivalent to Figure 6.4 when $\mu_i = \mu$, $1 \le i \le M$

and

$$X_2^*(n) = \frac{1}{\theta(1, M)\theta(2, M) \dots \theta(n, M)} = \binom{M + n - 1}{n}\mu^{-n}$$

$$(6.107)$$

then the probability that the virtual circuit (i.e. the server 2^*) holds n packets is given, from (6.27), as

$$P[W - n, n] = \frac{1}{G^*(W, 2)} X_1^*(W - n)X_2^*(n) \qquad (6.108)$$

or equivalently

$$\Pr[n_2^* = n] = P[W - n; n] = \frac{G^*(W - n, 1)}{G^*(W, 2)} X_2^*(n) \qquad (6.109)$$

where

$$G^*(k, 1) = X_1^*(k) \qquad (6.110)$$

and

$$G^*(k, 2) = \sum_{j=0}^{k} X_1^*(k - j)X_2^*(j) \qquad (6.111)$$

Thus the throughput of the virtual circuit is given by

$$\theta^*(W, 2) = \frac{G^*(W - 1, 2)}{G^*(W, 2)}$$

$$= \frac{\displaystyle\sum_{j=0}^{W-1} \mu^{-j}\binom{j + M - 1}{j}\Lambda(j + 1)}{\displaystyle\sum_{j=0}^{W} \mu^{-j}\binom{j + M - 1}{j}\Lambda(j)} \qquad (6.112)$$

Of course the same result could be derived from the original network with the $(M + 1)$ service centers. From (6.101) and the throughput formula we find that

$$\theta(W, M + 1) = \theta^*(W, 2) \qquad (6.113)$$

which is not unexpected. Needless to say, the average queue size, the end-to-end delay, and other quantities can be calculated in a similar manner either from the equivalent model of Fig. 6.5 or from the original system. The computation efforts that are required are essentially the same. The reader is referred to Schwartz (1980a; b) for further discussions of the flow control analysis based on these models.

REFERENCES

BARLOW, R. E. and PROSCHAN, F. (1967). "Mathematical Theory of Reliability", John Wiley & Sons Inc., New York.

BASKETT, F., CHANDY, K. M., MUNTZ, R. R. and PALACIOS, F. G (1975). Open, closed and mixed networks of queues with different classes of customers. *J. ACM.* **22** (2), pp. 248–260.

BHARAK-KUMAR, K. (1980). Discrete-time queueing systems and their networks. *IEEE Trans. on Communications*, **COM-28**, (2) pp. 260–263.

BUZEN, J. P. (1973). Computational algorithms for closed queueing networks with exponential servers. *CACM*, **16** (9), pp. 527–531.

CHANDY, K. M. AND SAUER, C. H. (1980). Computational algorithms for product form queueing networks. *CACM*, **23** (10), pp. 573–583.

CHANDY, K. M., HERZOG, U. and WOO, L. (1975). Parametric analysis of queueing networks. *IBM J. Research and Development*, **19** (1), pp. 36–42.

CHU, W. W. (1970). Buffer behavior for Poisson arrivals and multiple synchronous constant outputs. *IEEE Trans. on Computers*, **C-19** (6).

COX, D. R. and MILLER, H. D. (1965). "The Theory of Stochastic Processes", John Wiley & Sons Inc., New York, Chapters 2 and 5, pp. 22–75 and 203–251.

FELLER, W. (1971). "An Introduction to Probability Theory and Its Applications", Vol. 2, 2nd edition, John Wiley & Sons Inc., New York.

FISZ, M. (1963). "Probability Theory and Mathematical Statistics", 3rd edition, John Wiley & Sons Inc., New York.

GAVER, D. P. (1968). Diffusion approximations and models for certain congestion problems. *J., Appl. Prob.* **5**, pp. 607–623.

GAVER, D. P. (1971). Analysis of remote terminal backlogs under heavy demand conditions. *J. ACM.* **18**, (3), pp. 405–415.

GEIHS, K. and KOBAYASHI, H. (1980). Bounds on Buffer Overflow Probabilities in Communication Systems. Technical Report. Also to appear in *Performance Evaluation*, **2**, No. 3 (1982).

HSU, J. (1974). Buffer behavior with Poisson arrival and geometric output processes. *IEEE Trans. on Communications*, **COM-22**, (12), pp. 1940–1941.

HSU, J. and BURKE, P. J. (1976). Behavior of tandem buffers with geometric input and Markovian output. *IEEE Trans. on Communications*, **COM-24** (3), pp. 358–361.

JACKSON, J. R. (1963). Job shop-like queueing systems. *Management Sciences*, **10** (1), 131–142.

KINGMAN, J. F. C. (1962). On queues in heavy traffic. *J. Roy. Stat. Soc.*, **B, 24** pp. 383–392.

KINGMAN, J. F. C. (1964). A martingale inequality in the theory of queues. *Proc. Cambridge Philosophical Society*, **59**, pp. 359–361.

KINGMAN, J. F. C. (1965). The Heavy Traffic Approximation in the Theory of Queues. in Proc. Symp. on Congestion Theory, W. L. Smith and W. E. Wilkinson (Eds.), University of North Carolina Press, Chapter 6, 137–169.

KINGMAN, J. F. C.(1970). "Inequaliies in the theory of queues", *J. Roy. Stat. Soc.*, **32**, 102–110.

KLEINROCK, L. (1975). "Queueing Systems, Vol. I: Theory", John Wiley & Sons Inc., New York.

KLEINROCK, L. (1976). "Queueing Systems, Vol. II: Computer Applications", John Wiley & Sons Inc., New York.

KOBAYASHI, H. (1974a). Bounds for the Waiting Time in Queueing Systems. IBM Research Report RC4718, IBM Research Center, Yorktown Heights, New York 10598, February 1974.

KOBAYASHI, H. (1974b). Bounds for the waiting time in queueing systems. *In* "Computer Architectures and Networks", E. Gelenbe and R. Mahl (Eds.), North-Holland Publishing Company, pp. 263–274.

KOBAYASHI, H. (1974c). Application of the diffusion approximation to queueing networks I: equilibrium queue distributions, *J. ACM*, **21** (2), pp. 316–328.

KOBAYASHI, H. (1974d). Application of the diffusion approximation to queueing networks II: nonequilibrium distributions and applications to computer modeling. *J. ACM*, **21** (3), pp. 459–469.

KOBAYASHI, H. (1975). On Discrete-Time Processes in a Packetized Communication System. The ALOHA System Technical Report B75-78, University of Hawaii, Honolulu, HI96827, August 1975.

KOBAYASHI, H. (1976). A Computational Algorithm for Queue Distribution via the Polya Theory of Enumeration. IBM Research Report RC-6154, August 1976.

KOBAYASHI, H. (1978a). "Modeling and Analysis: An Introduction to System Performance Evaluation Methodology", Addison-Wesley. Reading, MA.

KOBAYASHI, H. (1978b). System design and performance analysis using analytic models. *In* "Current Trends in Programming Methodology, Vol. III: Software Modeling", K. M. Chandy and R. T. Yeh (Eds.), Prentice-Hall, Englewood Cliffs, NJ. pp. 72–114.

KOBAYASHI, H. and KONHEIM, A. G. (1977). Queueing models for computer communications system analysis. *IEEE Trans. on Communications*, **COM-25**. (1), pp. 2–29.

KOBAYASHI, H. and REISER, M. (1975). On Generalization of Job Routing Behavior in a Queueing Network Model. IBM Research Report RC-5252, February 1975.

KOBAYASHI, H., ONOZATO, Y. and HUYNH, D. (1977). An approximate method for design and analysis of an ALOHA System. *IEEE Trans. on Communications*, **COM-25** (1), pp. 148–157.

KONHEIM, A. G. (1975). An elementary solution of the queueing system G/G/1. *SIAM J. Computing*, **4** (4), pp. 540–545.

LAVENBERG, S. S. and REISER, M. (1979) Stationary Static Probabilities of Arrival Instants for Closed Queueing Network with Multiple Types of Customers. IBM Research Report RC-7592, IBM T. J. Watson Research Center, Yorktown Heights, April 1979. Also in *J. Appl. Prob.*, Dec. 1980.

LINDLEY, D. V. (1952). The theory of queues with a single server. *Proc. Cambridge Philosophical Society* **48**, pp. 277–289.

MIRASOL, N. M. (1963). The output of an M/G/∞ queueing system is Poisson *Operations Research*, **11**, pp. 282–284.

MORRISON, J. A. (1979). Two discrete-time queues in tandem. *IEEE Trans. on Communications*, **COM-27**, (3), pp. 563–573.

MORSE, P. M. and FESHBACH, H. (1953). "Method of Theoretical Physics" McGraw-Hill, New York, Part I, Chapter 6.

NEWELL, G. F. (1971). "Applications of Queueing Theory", Chapman and Hall London.

REISER, M. (1981). Mean-value analysis and convolution method for queue-dependent servers in closed queueing networks. *Performance Evaluation*, **1** (1). North-Holland Publishing Co, Amsterdam.

REISER, M. and KOBAYASHI, H. (1974). Accuracy of the diffusion approximation for some queueing systems. *IBM J. Research and Development*, **18** (2), pp 110–124.

REISER, M. and KOBAYASHI, H. (1975). Queueing networks with multiple closed chains: theory and computational algorithms. *IBM J. of Research and Development*, 19, (May), pp. 282–294.

REISER, M. and LAVENBERG, S. S. (1980). Mean value analysis of closed multichain queueing networks. *J. ACM*, **22** (April), pp 313–322.

ROSS, S. M. (1974). Bounds on the delay distribution in GI/G/1 queues. *J. Appl. Prob.*, **11**, pp. 417–421.

SCHWARTZ, M. (1977). "Computer Communication Network Design and Analysis", Prentice-Hall, Englewood Cliffs, NJ.

SCHWARTZ, M. (1980a). Routing and Flow Control in Data Networks. IBM Research Report RC8353, July 1980.

SCHWARTZ, M. (1980b). Performance Analysis of the SNA Virtual Route Pacing Control. IBM Research Report RC8490, September 1980.

SEVCIK, K. C. and MITRANI, I. (1979). The Distribution of Queueing Network States at Input and Output Instants. Proc. 4th Int. Symp. on Modeling and Performance Evaluation of Computer Systems, Vienna.

SMITH, W. L. (1953). On the distribution of queueing times. *Proc. Cambridge Philosophical Society*, **49**, pp. 449–461.

SOMMERFELD, A. (1949). "Partial Differential Equations in Physics", Academic Press, New York and London.

STONE, H. (1973). "Discrete Mathematical Structures and Their Applications", Science Research Associates.

SWEET, A. L. and HARDIN, J. C. (1970). Solutions for some diffusion processes with two barriers, *J. Appl. Prob.*, **7**, pp. 423–431.

WALD, A. (1947). "Sequential Analysis", John Wiley & Sons Inc., New York, pp. 157–160.

WYNER, A. D. (1974). On the probability of buffer overflow under an arbitrary bounded input-output distribution. *SIAM J. Appl. Math.*, **27** (4), pp. 544–570.

PART III

Mathematical Analysis of Combinatorial Algorithms

Robert Sedgewick

PREFACE

Computer programs as objects of study for mathematical analysis can be complicated and unsatisfying, or simple and elegant. The detailed study of the dynamic properties of computer programs, an intersection of fundamental old techniques from mathematical analysis with fundamental new techniques from computer science, is an interesting field of study around which a substantial body of knowledge has been built over the past fifteen years. This is a survey (partly) and tutorial (mostly) treatment of some of this work, presented at a level appropriate for both computer scientists and mathematicians.

The major reference work for the material described here is D. E. Knuth's *The Art of Computer Programming*, the first three volumes of which have been published (Knuth 1973a, 1973b, 1980). Knuth pioneered the art of detailed mathematical analysis of algorithms at the level considered here, and many of the derivations that we will consider in detail are taken from his work. In a sense, this could be viewed as an introduction to the serious mathematical material in Knuth's books, for the reader interested in learning what the books have to offer and interested in doing research in the area.

In some places, material contained here summarizes results which have appeared in the research literature. For example, most of the material in Chapter 12 comes from Yao (1976) and Knuth and Schonhage (1978), and some of the material in Chapters 10 and 11 comes from Sedgewick (1978b). Where possible, these notes are intended to supplement such material: the treatment here may seem sketchy because full details are presented in the papers.

Another important source for the preparation of these chapters was the notes for a graduate course in the mathematical analysis of algorithms introduced by Knuth at Stanford University in 1974, taught there by Knuth in 1976 and 1980 and by A. Yao in 1978, and taught by me at Brown University in 1977 and 1981. A significant part of these courses comprised detailed notes prepared by the graduate teaching assistants; these are quite well written and are invaluable as reference material.

A third influence on the material presented here has been the work of P. Flajolet at INRIA in France. Flajolet reintroduced me to the idea of avoiding laborious calculations with recurrences and sums for many problems by doing direct derivations with combinatorial generating functions, then using classical analysis to do asymptotics on the generating functions. I believe that this approach simplifies the analysis significantly for many problems and may have the potential to allow us to extend the range of algorithms that we can analyze; Flajolet and others are doing active research in this area.

The material is intended to be largely self-contained, but assumes that the reader has some familiarity with computer programming and with discrete mathematics. A broad range of material is presented in both domains, so every reader is likely to find things that are too elementary or too advanced. On the one hand, enough elementary mathematical material is included so that these notes may serve as an introduction to applicable mathematical analysis for a trained computer scientist: on the other hand, enough elementary algorithmic material is included so that they may serve as an introduction to combinatorial algorithms for the trained mathematician. My experience in teaching this material to a mix of computer scientists and mathematicians has been that the mathematicians are as challenged by the algorithms as the computer scientists are by the analysis.

Chapter 7 introduces the subject, with an outline of general techniques and a detailed example. Chapter 8 describes algorithms on trees and techniques for solving simple recurrences. Chapter 9 describes algorithms on permutations and introduces the use of generating functions. Chapter 10 introduces asymptotic methods through a detailed treatment of two particular algorithms. Chapter 11 extends the treatment of asymptotics in Chapter 10 by showing how methods from complex analysis must be used for many problems. Chapter 12 deals with analyses for a problem where several different simple algorithms and several different input models have been suggested: its purpose is to review much of the previous material and to illustrate the difficulty (and importance) of using the proper input model for many problems.

Acknowledgments

The generous support of IBM Belgium and the Université Libre de Bruxelles, which made possible the preparation of these notes and the lectures upon which they are based, is gratefully acknowledged. Also, some of this work was done under support from the National Science Foundation, Grant MCS80-17579 while the author was at Brown University.

Many people looked at an early draft of these notes and provided useful comments, including Trina Avery and Tom Freeman. In particular, Janet Incerpi carefully read and corrected several drafts.

7. Introduction

In this chapter we examine on a general level the basic approach espoused
by Knuth (1971) for the detailed mathematical analysis of algorithms. First
we consider the general motivations for algorithmic analysis, then we look
at the major components of a full analysis, then we analyze an algorithm of
fundamental practical importance, Quicksort, and then we discuss the
material to appear in later chapters.

7.1 WHY ANALYZE AN ALGORITHM?

There are several answers to this basic question, depending on context: the
intended use of the algorithm, the importance of the algorithm in relationship to others (from both practical and theoretical standpoints), and the
difficulty of analysis and accuracy of the answer required.

First, the most straightforward reason for analyzing an algorithm is to
discover its vital statistics in order to evaluate its suitability for various
applications or compare it with other algorithms. Generally, the vital statistics of interest are the primary resources of time and space, most often
time. Put simply, we are interested in determining how long an implementation of a particular algorithm will run on a particular computer, and how
much space it will require. The analysis generally is kept relatively independent of particular implementations, concentrating instead on deriving
results for essential characteristics of the algorithm which can be used to
estimate precisely true resource requirements on actual machines.

Occasionally, some expensive resource other than time or space is of
interest, and the focus of the analysis changed accordingly. For example, an
algorithm to drive a plotting device might be studied to determine the total
distance moved by the pen. Also, it is sometimes appropriate to combine
resources in the analysis. For example, an algorithm which uses a large

amount of memory may use much less time than an algorithm which gets by with very little memory. One way to compare algorithms in such situations is to analyze the product of their time and space requirements: this corresponds to using a "memory rental fee" as the resource to be studied.

The analysis of an algorithm can help one to understand it better, and can suggest informed improvements. The more complicated the algorithm, the more difficult the analysis, and algorithms tend to become shorter, simpler, and more elegant during the analysis process. More important, the careful scrutiny required for proper analysis often leads to more efficient and more correct implementations of algorithms. Analysis requires a far more complete understanding of an algorithm than merely producing a working implementation. Indeed, when the results of analytic and empirical studies agree, one becomes strongly convinced of the validity of the algorithm as well as of the correctness of the process of analysis.

Some algorithms are worth analyzing because their analysis can add to the body of mathematical tools available for mathematical analysis. Such algorithms may be of no practical interest, but may have properties similar to algorithms of practical interest which indicate that understanding them may help to someday understand more important methods. Unfortunately, results of this type are most often negative: many algorithms which seem to be very simple require extremely sophisticated mathematical machinery.

Many algorithms (some of intense practical interest, some of little or none) have a complex performance structure with properties of independent mathematical interest. The dynamic element brought to combinatorial problems by the analysis of algorithms leads to challenging, interesting mathematical problems which are worth studying in their own right.

7.2   GENERAL METHOD

The following general methodology is commonly used for the precise study of the performance of particular algorithms. This approach is a natural one and is very old, but it is generally attributed to Knuth (1971), whose books serve as witness to the utility of the method for fully understanding important algorithms.

The first step is to carefully implement the algorithm on a particular computer. We shall use the term *program* to describe such an implementation, so that one algorithm corresponds to many programs. This implementation not only provides a concrete object to study, but also can give useful empirical data to aid in or to check the analysis.

The implementation presumably is designed to make efficient use of expensive resources. The resources of primary interest must be identified so that the detailed analysis may be properly focused. The resource most often analyzed is the running time, so the steps below are outlined in terms of studying the running time.

The next step is to estimate the time required by each component instruction of the program. This can usually be done very precisely, depending on the characteristics of the computer system being used.

To determine the total running time of the program, it is necessary to study the branching structure of the program in order to express the frequency of execution of the component instructions in terms of unknown mathematical quantities. If the values of these quantities are known, then the running time of the entire program can be derived simply by multiplying the frequency and time requirements of each component instruction and adding these products.

The next step is to model the input to the program, to form a basis for the mathematical analysis. Often several different models are used for the same algorithm: initially a model is chosen to make the analysis as simple as possible; finally a model is chosen to reflect the actual situation in which the program is to be used.

The last step is to analyze the unknown quantities, assuming the modeled input. For average-case analysis, the quantities can be analyzed individually, then the averages can be multiplied by instruction times and added to give the running time of the whole program. For worst-case analysis, it is usually difficult to get an exact result for the whole program, and so an upper bound is often derived by multiplying worst-case values of the individual quantities by instruction times, then adding.

The average-case results can be compared with the empirical data to verify the implementation, the model, and the analysis.

7.3 AN EXAMPLE

To illustrate the methodology outlined above, results are sketched here for a particular algorithm of importance, the Quicksort sorting method. This analysis is covered in great detail elsewhere (Sedgewick, 1980 and 1977b), so only a very brief treatment will be given here.

First, an implementation of Quicksort in the PASCAL programming language follows:

```
var a: array [0 . .N] of integer;
procedure quicksort (l, r: integer);
    var v, t, i, j: integer;
    begin
```

if $r > 1$ **then**
  begin
  $v := a[r];\ i := 1 - 1;\ j := r;$
  repeat
    repeat $i : = i + 1$ **until** $a[i] > = v;$
    repeat $j := j - 1$ **until** $a[j] <= v;$
    $t := a[i];\ a[i] := a[j];\ a[j] := t;$
  until $j < i;$
  $a[j] := a[i];\ a[i] := a[r];\ a[r] := t;$
  quicksort (l, j);
  quicksort (i + 1, r)
  end
end;

This is a recursive program which sorts the numbers in an **array** $a[l:r]$ by partitioning it into two independent parts, then sorting those parts. The partitioning process puts the element that was in the last position in the array (the *partitioning element*) into its correct position, with all smaller elements before it and all larger elements after it. This is accomplished by maintaining two pointers, one scanning from the left, one from the right. The left pointer is incremented until an element larger than the partitioning element is found, the right pointer is decremented until an element smaller than the partitioning element is found. These two elements are exchanged, and the process continues until the pointers meet, which defines where the partitioning element is put. The call *quicksort(1,N)* will sort the array, provided that $a[0]$ is set to a value smaller than any other element in the array (in case, for example, the first partitioning element happened to be the smallest element).

The first step in the analysis is to estimate the resource requirements of individual instructions for this program. This is a straightforward step for any particular computer, and we will omit the details. For example, the "inner loop" instruction **repeat** $i: = j \div 1$ **until** $a[i] > = v$ translates, on most computers, to assembly language instructions like

$$\text{LOOP} \quad \text{INC} \quad \text{I, 1}$$
$$\text{CMP} \quad \text{V, A(I)}$$
$$\text{BL} \quad \text{LOOP}$$

This might require four time units (one for each memory reference).

The next step in the analysis is to assign variable names to the frequency of execution of the instructions in the program. Normally there are only a few true variables involved: the frequencies of execution of all the instructions can be expressed in terms of these few. Also, it is desirable to relate the variables to the algorithm itself, not any particular program. For Quicksort, there are three natural quantities involved:

A—the number of stages;
B—the number of exchanges; and
C—the number of comparisons.

On a typical computer, the total running time might be about $4C + 11B + 35A$. (The exact values of these coefficients depend on the assembly language program and properties of the machine being used; the values given above are typical.)

The input model for the analysis is to assume that the **array** *a* contains randomly ordered, distinct numbers. This is the most convenient to analyze; however, it is also possible to study this program under perhaps more realistic models allowing equal numbers (see Sedgewick 1977a).

The average-case analysis for this program involves defining and solving recurrence relations which mirror directly the recursive nature of the program. For example, if C_N is the average number of comparisons to sort *N* elements, we have $C_0 = C_1 = 0$ and

$$C_N = N + 1 + \frac{1}{N} \sum_{1 \le j \le N} (C_{j-1} + C_{N-j}) \quad \text{for} \quad N > 1$$

To get the total average number of comparisons, we add the number of comparisons for the first partitioning stage $(N + 1)$ to the number of comparisons used for the subfiles after partitioning. When the partitioning element is the *j*th largest (which occurs with probability $1/N$ for each $1 \le j \le N$), the subfiles after partitioning are of size $j - 1$ and $N - j$. For this to be valid, it is necessary to prove that the subfiles left after partitioning a random file are still random. Now the analysis is reduced to a mathematical problem which does not depend on properties of the program or the algorithm. This particular problem is not difficult to solve: first change *j* to $N - j + 1$ in the second part of the sum to get

$$C_N = (N + 1) + \frac{2}{N} \sum_{1 \le j \le N} C_{j-1}$$

Then multiply by *N* and subtract the same formula for $N - 1$ to eliminate the sum:

$$NC_N - (N - 1)C_{N-1} = 2N + 2C_{N-1}$$

Rearrange terms and divide by $N(N + 1)$ to get a simple recurrence

$$\frac{C_N}{N + 1} = \frac{C_{N-1}}{N} + \frac{2}{N + 1}$$

which holds for $N > 2$. This telescopes to a simple sum, giving the result

$$C_N = 2(N + 1)(H_{N+1} - \tfrac{4}{3})$$

where H_{N+1} is the $(N + 1)$st harmonic number $(\Sigma_{1 \leq k \leq N+1} 1/k)$. The average values of the other quantities can be derived in a similar manner.

This program can be improved in several ways to make it the sorting method of choice in many computing environments. A complete analysis can be carried out even for much more complicated improved versions, and expressions for the average running time can be derived which match closely observed empirical times (Sedgewick 1980, 1978a). Furthermore, the combinatorial mathematics involved in these analyses becomes quite interesting (see Sedgewick, 1977b).

7.4 PERSPECTIVE

The analysis above is in many ways an "ideal" methodology: not all algorithms can be as smoothly dealt with as this.

First, a full analysis like that above requires a fair amount of effort which should be reserved only for our most important algorithms. Most often, the parts of the methodology which are program-specific (dependent on the implementation) are skipped, to concentrate either on algorithm design, where rough estimates of the running time may suffice, or on the mathematical analysis, where the formulation and solution of the mathematical problem involved are of most interest. These are the areas involving the most significant intellectual challenge, and deserve the attention that they get. However, it is an unfortunate fact that as full an analysis as possible seems to be required to compare algorithms properly. Many researchers have been led astray by prematurely applying incomplete analyses.

In succeeding chapters, we will concentrate on techniques of mathematical analysis which are applicable to the study of the performance of algorithms. At the same time, we will survey some fundamental combinatorial algorithms, including several of practical importance. We will see that algorithms which seem to be quite simple can lead to quite intricate mathematical analyses, but that the analyses can uncover significant differences between algorithms which have direct bearing on the way they are used in practice.

The most serious problem in the analysis of most algorithms in common use on computers today is the formulation of proper models which realistically represent the input and which lead to manageable analysis problems.

Serious research in this seems to be required in several areas of application. However, there is a large class of *combinatorial* algorithms for which the model is very straightforward, as in sorting.

For the most part, algorithms of this type will be considered here. Many of these algorithms are of fundamental importance in a wide variety of computer applications, and so are deserving of the effort involved for detailed analysis. Furthermore, the input model leads immediately to mathematical problems of a combinatorial nature similar to those which arise in probability, so that classical methods of analysis provide a firm basis for their solution. The number of algorithms which have been studied in this way is steadily growing, and the analytic tools available are becoming increasingly better understood by computer scientists.

8. Trees

Trees are fundamental structures used in many practical algorithms, and it is important to understand their properties in order to be able to analyze these algorithms. In this section, we examine in detail several problems in analysis which relate to trees and to a fundamental tree search algorithm. These analyses not only provide results of practical interest, but also exhibit several fundamental techniques: for solving linear recurrences, for using generating functions and probability generating functions, and for using double and combinatorial generating functions.

The properties of trees have been studied for quite some time by combinatorial mathematicians, and many results are known. However, as we will see, there is a difference between viewing trees as static (combinatorial) objects and viewing them as dynamic objects, built by algorithms.

8.1 BINARY TREE SEARCH

The so-called "dictionary", "symbol table" or simply "search" problem is a fundamental one in computer science: a set of keys (perhaps with associated information) is to be organized so that efficient searches can be made for particular keys (or associated information). A *binary search tree* is an elementary structure commonly used for this problem: it consists of a root node containing a key and links to left and right subtrees which are defined in the same way. The left subtree contains all keys less than the key at the root, and the right subtree contains all keys greater. It is convenient to include an artificial "header" node with a key smaller than all other keys ($-\infty$) as the root of every binary tree. For example, the following binary tree contains the French words for the numbers 1 to 10. Such a tree can be

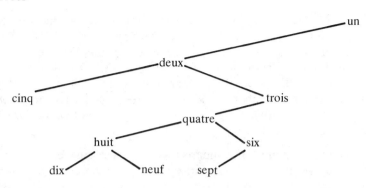

earched for a node with the value *v* using the following program:

```
: = head;
repeat
    if v < x ↑ .key then x: = x ↑ .left else x: = x ↑ .right
until v = x ↑ .key
```

f the key sought is not in the tree, this program will not work properly,
nce *x* will eventually be assigned to one of the null pointers at the bottom
f the tree. The program is easily modified to check for this case and to
sert a new node with the key sought if desired (see Knuth, 1973b;
edgewick 1983). For example, if the key *onze* were to be inserted, a new
ode would be created as the right son of *neuf*. Note that the nodes of the
ree can be printed out in order with the simple program:

```
rocedure print (x: link);
    begin
    if x < > nil then
        begin
        print (x ↑ .left);
        write (x ↑ .key);
        print (x ↑ .right)
        end
nd;
```

Many other useful operations are easily defined on binary search trees; see
Knuth (1973b).

There seem to be two quantities of interest in the analysis of the binary
ree search program: the number of nodes visited in a successful search and
he number of nodes visited in an unsuccessful search. These turn out to be
losely related. First, call the nodes of the tree which have keys *internal
odes*, and define imaginary nodes at the bottom called *external nodes*

(pointed to by null links). Note that any tree has exactly one more external node than internal node (this is trivial to prove by induction). The *internal path length* of the tree is defined to be the sum of the distances from the root to each internal node, and the *external path length* is defined analogously. Note that the external path length of any tree is the internal path length plus twice the number of nodes in the tree (again, trivial to prove by induction). Also, the quantities we want to analyze are obvious. For a tree with N nodes, if we define

C_N = Average number of comparisons for a successful search, and
C'_N = Average number of comparisons for an unsuccessful search then we have

$$C_N = \frac{\{\text{Internal path length}\}}{N} + 1$$

and

$$C'_N = \frac{\{\text{External path length}\}}{(N + 1)}$$

Then the relationship described above between internal and external path length implies that

$$(N + 1)C'_N = NC_N + N$$

Now, our notion of "average" must be more carefully defined. The above relationships hold for any given tree, if each internal (or external) node is equally likely to be sought: they are static properties of the tree. Our analysis must take into account the dynamic properties of the trees, since the way in which the keys are inserted can drastically affect the shape of the tree.

The tree above was created by inserting the keys in the order *un, deux, trois, quatre*, etc. If they are inserted in alphabetic order (*cinq, deux, dix, huit*, etc.) then a degenerate tree with a high path length results. If N nodes are inserted into an initially empty tree in order, then the resulting tree is a single string of nodes, connected by their right links, all with null left links. In other words, there is one internal node at distance i from the root for each $0 \le i < N$, so the internal path length is $\sum_{0 \le i < N} i = N(N - 1)/2$. Thus trees with quite different shapes can be built by the algorithm, and path lengths can vary widely. But, as we will see, it is *not* true that every possible tree is equally likely to result.

Knuth (1973b) gives a simple derivation that gives the average values of C_N and C'_N; we will do a more direct derivation later. The simple argument is to observe that the number of comparisons needed to find a key in the tree is exactly one greater than the number that was needed to insert it ince keys never move in the tree. Any particular key searched was the kth one inserted with probability $1/N$, so we have the recurrence

$$C_N = 1 + \frac{1}{N} \sum_{1 \le k \le N} C'_{k-1}$$

This is virtually the same as the recurrence that we had previously for Quicksort (in fact, the binary tree search algorithm is closely related to Quicksort). It can be solved by multiplying by N and subtracting the same equation for $(N - 1)$ to eliminate the sum.

$$(N + 1)C'_N - NC'_{N-1} = 2 + C'_{N-1}$$

Rearranging terms, we have

$$C'_N = C'_{N-1} + \frac{2}{N + 1}$$

which telescopes to the answer

$$C'_N = 2H_{N+1} - 2$$

which means that

$$C_N = 2\left(1 + \frac{1}{N}\right)H_{N+1} - \frac{2}{N} - 3$$

Now, H_N is about $\ln N$ (we will see how to say so more precisely in Chapter 10), so that, with N nodes in the tree, only about $2\ln N$ nodes need be examined for a typical search.

A fundamental point in this derivation is that we are "averaging" not over all trees, but over all possible orders of the keys inserted into the tree. (The reader should check that our assumptions are equivalent to this statement.) This "permutation" model is natural and realistic for this problem, but it is quite different from the assumption that all trees are equally likely to occur.

8.2 DIGRESSION: SOLVING FIRST-ORDER LINEAR RECURRENCES

Above, we took the recurrence

$$(N + 1)C'_N = (N + 1)C'_{N-1} + 2$$

and divided by $(N + 1)$ to get a recurrence which telescoped. For Quicksort, we had a more complicated recurrence

$$NC_N = (N + 1)C_{N-1} + 2N$$

which telescopes when divided by $N(N + 1)$. It turns out that it is always possible to transform recurrences of this nature into sums by telescoping. For example,

$$NC_N = (N - 2)C_{N-1} + 2N$$

telescopes when multiplied by $(N - 1)$, and

$$C_N = 2C_{N-1} + N$$

telescopes when divided by 2^N.

In general, the recurrence

$$C_N = X_N C_{N-1} + Y_N$$

telescopes when divided by $X_N X_{N-1} \ldots X_1$ (written $\prod_{N \geq k \geq 1} X_c$). (The reader may wish to check this formula on the examples above.) For some problems, the recurrence can be made to telescope by multiplying by $\prod_{K > N} X_K$, if it converges.

Solving recurrence relations (difference equations) in this way is analogous to solving differential equations by multiplying by an integrating factor and then integrating. The factor used for recurrence relations is sometimes called a "summation factor". Of course, for many problems, we may be left with a sum which is difficult to evaluate: we will study this in more detail later.

8.3   ELEMENTARY COMBINATORICS OF TREES

Trees as (static) objects have been intensely studied by combinatorial mathematicians, because they arise as natural models in many actual problems, and they have many interesting properties.

The most general kind of tree is a *free tree*: a set of N nodes and $N - 1$ edges connecting them together (this implies that there can be no cycles). If one of the nodes is designated as the *root*, then a free tree becomes an *oriented tree*, and if the order of the subtrees at every node is specified, we have an *ordered tree*. The first two trees below are equivalent as free trees but not as oriented trees; the second and third are equivalent as oriented

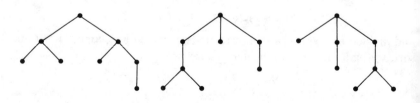

trees but not as ordered trees. A *binary tree* built by our algorithm is a special kind of ordered tree in which not only does each node have 0, 1 or 2 sons but also each son of a 1-son node is specified to be "left" or "right".

The most natural combinatorial question that arises for any defined tree structure is the enumeration problem: how many different trees are there? For some tree structures, this can be a difficult problem indeed; for binary trees it not only will shed some light on our algorithm, but also will illustrate another fundamental analytic tool.

If we let b_N be the number of different binary trees with N nodes, then the following recurrence relation holds:

$$b_N = \sum_{0 \le k < N} b_k b_{N-1-k}, \qquad N > 0$$

It is convenient to define $b_0 = 1$, $b_N = 0$ for all N negative, and make the recurrence hold for all N:

$$b_N = \sum_{0 \le k < N} b_k b_{N-1-k} + \delta_{N0}$$

(Here δ_{N0} is 1 for $N = 0$, 0 otherwise.) This can be checked against the small values in the table below:

N	0	1	2	3	4
b_N	1	1	2	5	14

This recurrence is much more complicated than those we have seen before, and requires more powerful tools. Define the *generating function*

$$B(z) = \sum_{N \ge 0} b_N z^N$$

where z is some artificial variable. This function describes the entire sequence succinctly. Multiplying both sides of the recurrence by z^N and summing on N, we have

$$B(z) = \sum_{N \ge 0} \sum_{0 \le k < N} b_k b_{N-1-k} z^N + 1$$

$$= \sum_{k \ge 0} \sum_{N > k} b_k b_{N-1-k} z^N + 1 \text{ (interchange order)}$$

$$= \sum_{k \ge 0} \sum_{N \ge 0} b_k b_N z^{N+k+1} + 1 \text{ (change } N \text{ to } N + k + 1)$$

$$= z \sum_{k \ge 0} b_k z^k \sum_{N \ge 0} b_N z^N + 1$$

$$B(z) = zB(z)^2 + 1$$

(A double sum of this type is called a *convolution*; it arises whenever two power series are multiplied.) This formula for $B(z)$ can be solved with the quadratic equation

$$B(z) = \frac{1}{2z}(1 \pm \sqrt{1 - 4z})$$

To solve for b_N we need to expand back to power series. This is easily done with the binomial theorem:

$$(1 - 4z)^{1/2} = \sum_{N \geq 0} \binom{\frac{1}{2}}{N}(-4z)^N$$

Thus, $B(z) = 1/2z(\pm (1 - 2z + \ldots))$ and we must choose the root with the negative sign in order for $B(0)$ to be defined. This equation can be used to derive b_N in familiar terms: the binomial coefficient $\binom{x}{N}$ is defined to be

$$\prod_{1 \leq j \leq N} \frac{x - j + 1}{N - j + 1}$$

This is the familiar $x!/N!(x - N)!$ for integer x, but the more general definition allows the binomial theorem to be used for noninteger exponents. This leads to

$$B(z) = \sum_{N \geq 0} \binom{\frac{1}{2}}{N + 1}(-1)^N 2^{2N+1} z^N$$

so we can set coefficients equal to get

$$b_N = \binom{\frac{1}{2}}{N + 1}(-1)^N 2^{2N+1} = \frac{1}{N + 1}\binom{2N}{N}$$

These numbers are called the Catalan numbers, which appear very frequently in combinatorics. In Chapter 10 we will see how to show that the approximate value is $b_N \approx 4^N/N\sqrt{\pi N}$.

There are many more permutations on N objects than binary trees of N items; therefore, many different permutations give rise to the same tree in the binary tree search algorithm. But not all trees are equally likely: as we shall see, some trees (fortunately the more balanced ones) appear much more frequently than others.

8.4 GENERATING FUNCTION SOLUTION OF RECURRENCES

Generating functions provide a mechanical method for solving many recurrence relations, although some facility for manipulating power series is required. For example, a direct solution for the binary tree search recurrence

$$(N + 1) C'_N = 2N + \sum_{0 \leq k < N} C'_k$$

can be derived as follows. Define $C(z) = \sum_{N \geq 0} C'_N z^{N+1}$, multiply both sides of the equation by z^N and sum on N to get

$$C'(z) = 2 \sum_{N \geq 0} Nz^N + \sum_{N \geq 0} \sum_{0 \leq k \leq N} C'_k z^N$$

The double sum is a convolution, as before, so

$$C'(z) = 2 \sum_{N \geq 0} Nz^N + \sum_{k \geq 0} C'_k z^{k+1} \sum_{N \geq 0} z^N$$

But

$$\sum_{N \geq 0} z^N = \frac{1}{1-z}$$

is elementary, and differentiating both sides gives

$$\sum_{N \geq 0} Nz^{N-1} = \frac{1}{(1-z)^2}$$

so we have a differential equation on the generating function

$$C'(z) = \frac{2z}{(1-z)^2} + \frac{C(z)}{1-z}$$

The solution to this differential equation is

$$\rho C(z) = \int \rho \frac{2z}{(1-z)^2} \, dz \quad \text{where} \quad \rho = \exp\left(-\int \frac{1}{1-z} \, dz\right)$$

Carrying out the calculation gives $\rho = 1 - z$ and

$$C(z) = \frac{2}{1-z} \ln \frac{1}{1-z} - \frac{2z}{1-z}$$

We know from the above that

$$\frac{1}{1-z} = \sum_{N \geq 1} z^N$$

Integrating this gives

$$\ln \frac{1}{1-z} = \sum_{N \geq 1} \frac{z^N}{N}$$

and multiplying both sides by $1/(1 - z)$ gives

$$\frac{1}{1 - z} \ln \frac{1}{1 - z} = \sum_{N \geq 1} H_N z^N$$

Setting coefficients of z^N equal gives the solution

$$C'_N = 2H_{N+1} - 2$$

as before.

For some problems, an explicit formula for the generating function can be difficult to derive. For others, the expansion back to power series can present the main obstacle. However, this general method can be relied on to produce a solution for many recurrences, if a solution is available.

8.5 UNSUCCESSFUL SEARCH IN BINARY SEARCH TREES

The above derivation shows how to find the average cost of an unsuccessful search in a binary tree, but it cannot be extended to find more information about the distribution of this quantity (for example, the variance). In this section, we examine a detailed direct analysis for this problem.

As above, we assume that a random permutation of N elements is used to build a binary search tree. Let $P_{Nk} \equiv \{$Probability that the last insertion takes k steps$\}$. Then the average unsuccessful search time is given by

$$\sum_k k P_{Nk}$$

and the variance by

$$\sum_k (k - C'_N)^2 P_{Nk} = \sum_k k^2 P_{Nk} - C'^2_N$$

To calculate P_{Nk} directly, the first step is to use a combinatorial argument based on permutations to set up a recurrence relation. This will be a general topic of the next section; we will postpone discussion of this particular argument until then. The recurrence which results is

$$N P_{Nk} = 2P_{(N-1)(k-1)} + (N - 2)P_{(N-1)k}$$

If we use the generating function

$$P_N(z) = \sum_{k \geq 0} P_{Nk} z^k$$

we get, after multiplying by z^k and summing,

$$N P_N(z) = (2z + N - 2)P_{N-1}(z)$$

which telescopes immediately to

$$P_N(z) = \prod_{2 \le j \le N} \frac{2z + j - 2}{j}$$

This complicated explicit formula for our generating function is actually well known: it turns out that

$$P_{Nk} = \frac{1}{N!} \begin{bmatrix} N - 1 \\ k \end{bmatrix} 2^k$$

where the brackets indicate Stirling numbers of the first kind. The mean and variance can then be calculated directly (although some facility with sums involving Stirling numbers is necessary). Fortunately, there is a much easier way.

8.6 PROBABILITY GENERATING FUNCTIONS

When generating functions are used to manipulate probabilities, they have several simple properties which make the calculation of the average, variance, and other moments easy. If

$$P(z) = \sum_{k \ge 0} p_k z^k$$

where $\{p_k\}$ is a sequence of probabilities (positive values which sum to 1), then we can exploit the following facts:

(i) $P(1) = \sum_{k \ge 0} p_k = 1$;
(ii) the average, defined to be $\sum_{k \ge 0} k p_k$, is simply $P'(1)$;
(iii) the variance is $P''(1) + P'(1)) - P'(1)^2$; and
(iv) the average and variance of the distribution represented by the product of two probability generating functions is the sum of the individual averages and variances, because if $R(z) = P(z)Q(z)$ and $P(1) = Q(1) = 1$ then $R'(1) = P'(1) + Q'(1)$ and $R''(1) = P''(1) + Q''(1) + 2P'(1)Q'(1)$.

These four properties apply directly to our problem, because the generating function $P_N(z)$ is the product of a number of very simple probability generating functions

$$f_j(z) = \frac{j - 2}{j} + \frac{2}{j} z$$

(note that $f_j(1) = 1$). Therefore, we need only compute moments for these simple functions and then sum. We have

$$f_j'(1) = \frac{2}{j} \quad \text{and} \quad f_j''(1) = 0$$

so

$$P_N'(1) = \sum_{2 \le j \le N} \frac{2}{j} = 2H_N - 2$$

as before, and

$$P_N''(1) + P_N'(1) - P_N'(1)^2 = \sum_{2 \le j \le N} \frac{2}{j} - \frac{4}{j^2} = 2H_N - 4H_N^{(2)} + 2$$

Probability generating functions provide a very convenient way to calculate the mean and variance of the number of steps required for an unsuccessful search.

Two more useful properties of probability generating functions can help to directly formulate recurrence relations involving them:

(v) if $X(z)$ and $Y(z)$ are probability generating functions for independent random variables X and Y, then $X(z)Y(z)$ is the probability generating function for X + Y; and

(vi) if $X_1(z)$ and $X_2(z)$ are the probability generating functions for a random variable X, conditional on whether or not some independent event happens (with probability p), then the unconditional probability generating function for X is $pX_1(z) + (1 - p)X_2(z)$. For example property (v) can be used to give a direct argument for the equation

$$P_N(z) = \left(\frac{N - 2}{N} + \frac{2z}{N} \right) P_{N-1}(z)$$

for unsuccessful search: with probability $2/N$, the last two elements are on the same search path, contributing 1 to the total cost of an unsuccessful search; otherwise (with probability $(N - 2)/N$) the contribution is 0.

8.7   SUCCESSFUL SEARCH IN BINARY TREES

More detailed analyses of successful search can be carried out using the methods of the previous section in conjunction with the fundamental relationship between internal and external path length.

Another quantity of interest is the *total* internal path length: the total cost of building the tree. If we let $q_{Nk} = Pr \{k$ is the total internal path length of a tree built from a random permutation of N elements$\}$ then properties (v) and (vi) above lead directly to the recurrence

$$q_N(z) = \frac{z^{N-1}}{N} \sum_{1 \le j \le N} q_{j-1}(z) q_{N-j}(z)$$

on the generating function

$$q_N(z) = \sum_{k \ge 0} q_{Nk} z^k$$

This recurrence is difficult to solve explicitly for q_{Nk}, but differentiating and evaluating at $z = 1$ gives the familiar recurrence

$$q'_N(1) = N - 1 + \frac{1}{N} \sum_{1 \le j \le N} (q'_{j-1}(1) + q'_{N-j}(1))$$

for the average. The variance can be derived in a similar way.

8.8 INTERNAL PATH LENGTH IN STATIC TREES

We have seen that there are many fewer binary trees on N elements than permutations, and that the mapping from permutations to trees effected by the binary tree search algorithm is such that the average internal path length is $2(N + 1)H_{N+1} - 4N - 2$. In this section we analyze the internal path length of binary trees in the static model, where each tree is considered equally likely.

If we define q_{Nk} to be the probability that k is the total internal path length, then the same argument as before gives a recurrence relation on the generating function $q_N(z) = \sum_{k \ge 0} q_{Nk} z^k$:

$$q_N(z) = z^{N-1} \sum_{1 \le j \le N} \frac{\dfrac{1}{j}\dbinom{2j-2}{j-1} \dfrac{1}{N-j+1}\dbinom{2N-2j}{N-j}}{\dfrac{1}{N+1}\dbinom{2N}{N}} q_{j-1}(z) q_{N-j}(z)$$

(Before, the probability that the subtrees were of size $j - 1$ and $N - j$ was always $1/N$; here it is a complicated-looking function derived from the Catalan numbers in a simple way.) As before, this recurrence is hard to solve for q_{Nk}. Here, it is even difficult to find $q'_N(1)$: the resulting recurrence requires generating functions. It will be slightly more convenient to work with the number of trees of size N with internal path length k,

$$Q_{Nk} \equiv \frac{1}{N+1}\binom{2N}{N} q_{Nk}$$

which satisfies (from the above recurrence)

$$\sum_{k \geq 0} Q_{Nk} z^k = z^{N-1} \sum_{1 \leq j \leq N} \sum_{r \geq 0} Q_{(j-1)r} z^r \sum_{s \geq 0} Q_{(N-j)s} z^s$$

Since we will need a generating function of N later, it is convenient to include N right away in a *double generating function*:

$$Q(w, z) = \sum_{N \geq 0} \sum_{k \geq 0} Q_{Nk} w^N z^k$$

Multiplying the above by w^N and summing on N gives

$$Q(w, z) = \sum_{N \geq 1} \sum_{1 \leq j \leq N} \sum_{r \geq 0} Q_{(j-1)r} z^r \sum_{s \geq 0} Q_{(N-j)s} z^s w^N z^{N-1} + 1$$

$$= w \sum_{j \geq 0} \sum_{r \geq 0} Q_{jr} z^r (wz)^j \sum_{N \geq j} \sum_{s \geq 0} Q_{(N-j)s} z^s (wz)^{N-j} + 1$$

$$= w \sum_{j \geq 0} \sum_{r \geq 0} Q_{jr} z^r (wz)^j \sum_{N \geq 0} \sum_{s \geq 0} Q_{Ns} z^s (wz)^N + 1$$

$$= wQ(wz, z)^2 + 1$$

Now we can differentiate this with respect to z and evaluate it at 1 to get the average. Note that

$$Q(w, 1) = \sum_{N \geq 0} \sum_{k \geq 0} Q_{Nk} w^N$$

$$= \sum_{N \geq 0} \frac{1}{N+1} \binom{2N}{N} w^N = B(w),$$

the generating function for the Catalan numbers, and

$$q(w) \equiv \frac{\partial}{\partial z} Q(w, z) \Big|_{z=1}$$

$$= \sum_{N \geq 0} \sum_{k \geq 0} k Q_{Nk} w^N$$

$$= \sum_{N \geq 0} \frac{1}{N+1} \binom{2N}{N} q'_N(1) w^N$$

is the generating function for the total internal path length. Differentiating both sides of our equation for $Q(w, z)$ with respect to z and evaluating at $z = 1$ gives

$$q(w) = 2wB(w)(q(w) + wB'(w))$$

which has the solution

$$q(w) = \frac{1}{1 - 4w} - \frac{1}{w} \left(\frac{1 - w}{\sqrt{1 - 4w}} - 1 \right)$$

and gives the eventual result

$$q'_N(1) = \frac{4^N(N+1)}{\binom{2N}{N}} - 3N - 1 \approx N\sqrt{\pi N} - 3N$$

Notice that this is substantially larger, for large N, than for the dynamic case: the average internal path length for binary search trees is proportional to $N\log N$, not $N\sqrt{N}$. These results show that there are many degenerate trees which are very unbalanced, but these trees are built only rarely by the binary tree search algorithm.

8.9 COMBINATORIAL GENERATING FUNCTIONS

The derivation above includes some quite complicated manipulations with triple sums to prove the simple formula

$$Q(w,z) = wQ(wz,z)^2 + 1$$

It is reasonable to ask whether there might be a simpler proof of this formula. Fortunately there is, using a direct argument based upon the generating function itself.

Our approach to this point has been to treat probabilistic analyses as counting problems; then derive recurrence relations for the counting problems; then discover an implied relationship between associated generating functions, and then use analytic techniques to learn about the generating function from this relationship. However, for many, if not most, problems amenable to solution by this approach, it turns out to be easy to derive the relationship on the generating function directly. The key to this new approach is to use the combinatorial object being analyzed as the index of summation for the generating function rather than the parameters of the analysis as we have been using. For example, in the derivation above we were working with

$$Q(w,z) = \sum_{N \geq 0} \sum_{k \geq 0} Q_{Nk} w^N z^k$$

where Q_{Nk} is the number of trees with N nodes and internal path length k. This may be expressed equivalently as

$$Q(w,z) = \sum_{\text{all trees } T} w^{|T|} z^{ipl(T)}$$

where $|T|$ is the number of nodes in T and $ipl(T)$ is the internal path length of T. Now, we can change this into a double sum, because any nonempty tree T consists of left and right subtrees, trees T_L and T_R, joined together by a root node. This leads immediately to

$$Q(w, z) = \sum_{\text{all trees } T_L} \sum_{\text{all trees } T_R} w^{|T_L + T_R + 1|} z^{ipl(T_L) + ipl(T_R) + |T_L| + |T_R|} + 1$$

(The "$+1$" takes into account the case when T is empty). Note that the number of nodes in a tree is one plus the number of nodes in its subtrees, and the internal path length of a tree is the sum of the internal path lengths of its subtrees plus one for each node in the subtrees. Now, this double sum is easily rearranged to make two independent sums:

$$Q(w, z) = w \sum_{\text{all trees } T_L} (wz)^{|T_L|} z^{ipl(T_L)} \sum_{\text{all trees } T_R} (wz)^{|T_R|} z^{ipl(T_R)} + 1$$
$$= wQ(wz, z)^2 + 1$$

as before. The reader may wish to study this example carefully to appreciate both its simplicity and its subtleties.

For another example, consider the application of combinatorial generating functions to find the total internal path length of binary search trees. Here, the combinatorial objects of interest are permutations, so we start with the generating function

$$Q(w, z) = \sum_{\text{all perms } P} w^{|P|} \frac{z^{ipl(P)}}{|P|!}$$

Here $ipl(P)$ denotes the internal path length of the binary search tree constructed when the elements of P are inserted into an initially empty tree using the standard algorithm. Note that, in this generating function, we need to divide by $|P|!$. One reason for this is that there are many more permutations than trees and the function would not converge otherwise. Another reason that dividing by $|P|!$ is convenient is that it makes $Q(w, z)$ a probability generating function: we have the equivalent expression

$$Q(w, z) = \sum_{N \geq 0} \sum_{k \geq 0} P_{Nk} w^N z^k$$

where P_{Nk} is the probability that k is the internal path length of a tree built from a random permutation of N elements. As above, we can directly derive a functional equation by splitting the combinatorial definition into a double sum. Given two permutations P_L, P_R, we can create

$$\binom{|P_L| + |P_R|}{|P_L|}$$

permutations of size $|P_L| + |P_R| + 1$ all of which led to the same binary search tree by (i) adding $|P_L| + 1$ to each element of P_R; (ii) intermixing

P_L and P_R in all possible ways; and (iii) prefixing each permutation so obtained by $|P_L| + 1$. This leads to the following double sum representation for $Q(w, z)$:

$$\sum_{\text{all perms } P_L} \sum_{\text{all perms } P_R} \binom{|P_L| + |P_R|}{|P_L|} \frac{w^{|P_L|+|P_R|+1} z^{ipl(P_L)+ipl(P_R)+|P_L|+|P_R|}}{(|P_L| + |P_R| + 1)!} + 1$$

(The permutation of size 0 can not be split in this way: this accounts for the "$+1$".) This equation is somewhat more complicated than the one above, but it can be simplified by differentiating with respect to w first, then rearranging terms as above. If $Q_w(w, z)$ denotes the derivative of $Q(w, z)$ with respect to w, we have

$$Q_w(w, z) = \sum_{\text{all perms } P_L} \frac{(wz)^{|P_L|} z^{ipl(P_L)}}{|P_L|!} \sum_{\text{all perms } P_R} \frac{(wz)^{|P_R|} z^{ipl(P_R)}}{|P_R|!}$$

$$= Q(wz, z)^2$$

Now if we differentiate this with respect to z and evaluate at $z = 1$, then we get the differential equation for the average internal path length that we solved in the previous chapter.

8.10   ADVANCED TREE ALGORITHMS

The elementary binary tree search algorithm has the undesirable property of having a very bad worst case. Several more advanced structures have been developed to deal with this and other problems: the analysis of these structures leads to a wealth of interesting mathematical problems.

One type of solution uses *radix search trees*. These structures are built using individual bits of the keys, rather than comparing keys as entities. The worst-case performance is proportional to the number of bits in the keys. The analysis of the average-case performance for this method requires advanced techniques that we will study in a later section.

A second type of solution uses *balanced binary trees*. These structures are built with algorithms that do local transformations on binary trees during an insertion, to prevent them from getting too far out of balance. The worst case for any search is proportional to $\log N$. Properties and implementations of these algorithms are discussed in some detail in Guibas and Sedgewick (1978). The mathematical questions involved in studying these trees are interesting, but difficult: very few results have been derived. For example, for some of the algorithms, the difference between the static and the dynamic structures is even more pronounced than above. The algorithms achieve balance by allowing more keys per node, so that, for example, 3-nodes (with two keys and three sons) and 4-nodes (with three keys and four sons) are allowed. Enumerating all such trees is difficult enough (Odlyzko, 1979), but for some algorithms, there are trees which cannot even be constructed: the enumeration problem for the dynamic case is unsolved, let alone the path length problem.

The unsolved mathematical problems on trees are not restricted to advanced structures. For example, the average height (the length of the longest path from the root to an external node) has only recently been derived for the static model (Flajolet and Odlyzko, 1980, and the full answer for the dynamic model is still not known (Robson, 1979).

9 Permutations

Combinatorial algorithms deal only with the relative order of a linear array of N elements and so can be thought of as operating on the numbers 1 to N in some order. Such an ordering is called a *permutation*, and is a well-defined combinatorial object with a wealth of interesting properties. In the previous chapter, we analyzed an algorithm which transforms permutations into trees; in this chapter, we will look at the analysis of more algorithms on permutations, the properties of permutations that arise in these analyses, and some more tools (evaluation of finite sums involving harmonic numbers and binomial coefficients) for use in such analyses. The algorithms that we will study are much simpler than those in the previous section, but we will see that even simple properties of permutations can be difficult to analyze.

9.1 FINDING THE MINIMUM

The trivial algorithm for finding the minimum element in an array may be implemented as follows:

v: $= \infty$;

for i: $= 1$ **to** N **do if** $A[i] < v$ **then** v: $= A[i]$;

The running time of this algorithm, is proportional to $c_1 N + c_2 A + c_3$, where c_1, c_2, c_3 are appropriate constants and A (the only "variable" involved) is the number of times $v := A[i]$ is executed. This is the number of *left-to-right minima* of the permutation: the number of new minimum values encountered when scanning from left to right.

For example, the permutation

$$3 \quad 8 \quad 4 \quad \mathbf{2} \quad 5 \quad 9 \quad 6 \quad \mathbf{1} \quad 7$$

has three left-to-right minima (in bold face). As we have before, we will

151

assume all $N!$ permutations to be equally likely as input and define the probabilities

$$P_{Nk} \equiv Pr\{A = k\} = \frac{\{\text{number of perms for which } A = k\}}{N!}$$

The method of solution is to set up a recurrence relation on P_{Nk} by writing the permutations in an appropriate order, in this case sorted by their last element, as in the example below for $N = 4$.

1 2 3 4	1 2 4 3	1 3 4 2	2 3 4 1
2 1 3 4	2 1 4 3	3 1 4 2	3 2 4 1
1 3 2 4	1 4 2 3	1 4 3 2	2 4 3 1
3 1 2 4	4 1 2 3	4 1 3 2	4 2 3 1
2 3 1 4	2 4 1 3	3 4 1 2	3 4 2 1
3 2 1 4	4 2 1 3	4 3 1 2	4 3 2 1

Writing the permutations down in this way makes it plain that the last element does not affect the number of left-to-right minima unless it is the smallest. More precisely, every permutation of $N - 1$ elements with k left-to-right minima corresponds to $(N - 1)$ permutations of N elements with k left-to-right minima and 1 permutation of N elements with $(k + 1)$ left-to-right minima. This leads directly to the recurrence

$$N!P_{Nk} = (N - 1)(N - 1)!P_{(N-1)k} + (N - 1)!P_{(N-1)(k-1)}$$

$$P_{Nk} = \left(1 - \frac{1}{N}\right)P_{(N-1)k} + \frac{1}{N}P_{(N-1)(k-1)}$$

In terms of the probability generating function $P_N(z) = \Sigma_{k \geq 0}P_{Nk}z^k$, this is

$$P_N(z) = \frac{z + N - 1}{N}P_{N-1}(z)$$

This formula could be derived directly, as in Chapter 8, with the argument that the last element independently contributes 1 to the number of left-to-right minima with probability $1/N$. Also, as in Chapter 8, we can find the mean and variance of the number of left-to-right minima by summing the means and variance from the simple probability generating functions $(z + k - 1)/k$, with the eventual result that the mean is $H_N - 1$ with variance $H_N - H_N^{(2)}$.

This problem can also be dealt with using the "combinatorial" double generating function technique introduced in the previous chapter. While the derivation is certainly not simpler than the one given above, it is very instructive and will prepare us well for more difficult problems that cannot be handled easily with the elementary techniques used above. As before, the combinatorial objects of interest are permutations, so we start with the generating function

$$B(w, z) = \sum_{\text{all perms } P} w^{|P|} \frac{z^{lrm(P)}}{|P|!}$$

Here $lrm(P)$ denotes the number of left-to-right minima in the permutation P and an alternate representation for the generating function is

$$B(w, z) = \sum_{N \geq 0} \sum_{k \geq 0} P_{Nk} w^N z^k$$

where P_{Nk} is the probability that a random permutation of N elements has k left-to-right minima. As before, we can directly derive a functional equation from the combinatorial definition. Given a permutation P_1, we can create $|P_1| + 1$ permutations of size $|P_1| + 1$, one of which ends in 1 (and so has one more left-to-right minimum than P_1), and $|P_1|$ of which do not end in 1 (and so have the same number of left-to-right minima as P_1). This leads to the formulation

$$B(w, z) = \sum_{\text{all perms } P_1} \frac{w^{|P_1|+1} z^{lrm(P_1)+1}}{(|P_1| + 1)!} + \sum_{\text{all perms } P_1} \frac{|P_1| w^{|P_1|+1} z^{lrm(P_1)}}{(|P_1| + 1)!}$$

Differentiating with respect to w, we have

$$B_w(w, z) = \sum_{\text{all perms } P_1} \frac{w^{|P_1|} z^{lrm(P_1)+1}}{|P_1|!} + \sum_{\text{all perms } P_1} \frac{w^{|P_1|} z^{lrm(P_1)}}{(|P_1| - 1)!}$$

$$= zB(w, z) + wB_w(w, z)$$

Solving for $B_w(w, z)$, we get a simple first-order differential equation

$$B_w(w, z) = \frac{z}{1 - w} B(w, z)$$

which has the solution

$$B(w, z) = \frac{1}{(1 - w)^z}$$

(since $B(0, 0) = 1$). Differentiating with respect to z, we have

$$B_z(w, z) = \ln \frac{1}{1 - w} e^{z \ln(1/1 - w)}$$

Now, evaluating at $z = 1$ gives the result

$$B_z(w, 1) = \frac{1}{1 - w} \ln \frac{1}{1 - w}$$

the generating function for the harmonic numbers, as expected.

9.2 *IN-SITU* PERMUTATION

A direct use for permutations is to specify how to rearrange elements in an array. For example, the permutation

$$3 \quad 8 \quad 4 \quad 2 \quad 5 \quad 9 \quad 6 \quad 7$$

could direct that the array

$$E \quad A \quad S \quad R \quad M \quad L \quad D \quad C \quad B$$

should be rearranged by putting the third element in position 1, the eighth element in position 2, etc. to leave

$$S \quad C \quad R \quad A \quad M \quad B \quad L \quad E \quad D$$

If the permutation is in an array $P[1 : N]$, the array in $A[1 : N]$ and an output array $B[1 : N]$ is available, the program is trivial:

for $i := 1$ **to** N **do** $B[i] := A[P[i]]$;

In some situations, the B array might not be available, and the A array needs to be permuted "in place". A straightforward way to do this is to start by saying $A[1]$ in a register, replace it by $A[P[1]]$, set j to $P[1]$, and continue until $P[j]$ becomes 1, when $A[j]$ can be set to the saved value. This process is then repeated for each element not yet moved, so it requires a bit array to mark those elements which have been moved, as in the following implementation:

```
for j := 1 to N do
  if not m[j] then
    begin
    s := j; z := A[j]; t := P[j];
    while t <> j do
      begin m[s] := true; A[s] := A[t]; s := t; t := P[s] end;
    A[s] := z; m[s] := true;
    end;
```

In our example, first $A[3] = S$ is moved to position 1, then $A[4] = R$ is moved to position 3, etc., to leave

$$S \quad C \quad R \quad A \quad M \quad L \quad D \quad E \quad B$$

Next, the M is marked (and not moved) and then B, D and L are marked and moved.

This is not truly an "in-place" algorithm because N extra bits are needed for the marks: we will see how to eliminate the "mark" bits later. The only "variable" in the running time of this program is the number of times the **if** statement succeeds, the number of *cycles* in the permutation. The analysis of this quantity turns out to be simple because of a combinatorial correspondence between permutations. A permutation can be defined by writing out its cycles: the example permutation above can be written

$$(1\ 3\ 4\ 2\ 8)\ (5)\ (6\ 9\ 7)$$

which means that $A[3]$ is to be moved to position 1, then $A[4]$ is to be moved to position 3, etc. Since there are several ways to write the same permutation in this "cycle" notation (for example, (976)(5)(28134) is another for the permutation above), it is convenient to define a canonical representation: for each cycle, write the smallest element in the cycle first (call this the *leader*), then write the cycles in decreasing order of their leaders:

$$(6\ 9\ 7)\ (5)\ (1\ 3\ 4\ 2\ 8)$$

But now the parentheses are no longer needed, and we have a one-to-one correspondence with another permutation:

$$6\ 9\ 7\ 5\ 1\ 3\ 4\ 2\ 8$$

In fact, it is the left-to-right minima that determine where cycles begin and end. This means that we have already completed the analysis, and the average and variance for the number of cycles is the same as for the number of left-to-right minima, H_N and $H_N - H_N^{(2)}$.

To eliminate the "mark" bits, we can decide to permute each cycle only when its leader is encountered, as follows:

```
for j := 1 to N do
      begin
      k := P[j]; while k > j do k := P[k]
      if k = j then permute cycle;
      end;
```

Now the analysis must also include a quantity (call it B) that counts the iterations of this **while** loop, the "distance" from each element to a smaller one in the same cycle. This is most easily counted in the permutation describing the cycle structure: in our example, the following table gives the contribution to this quantity due to each element:

$$6\ 9\ 7\ 5\ 1\ 3\ 4\ 2\ 8$$
$$2\ 0\ 0\ 0\ 4\ 1\ 0\ 1\ 0$$

That is, when the 3 is encountered, one more element must be examined (the 4) before a smaller one in the same cycle (the 2) is encountered (which bears witness to the fact that 3 is not the cycle leader).

An easy way to analyze this quantity is to expand the table above to a two-dimensional array in which the ith row has 1s in positions corresponding to "right-to-left" minima in the permutation defined by considering only the first i positions. In our example, this table is

```
6  9  7  5  1  3  4  2  8
0  0  0  0  1  0  0  1  1
0  0  0  0  1  0  0  1
0  0  0  0  1  1  1
0  0  0  0  1  1
0  0  0  0  1
0  0  0  1
1  0  1
1  1
1
```

If we ignore the rightmost 1s, the columns in this table add up to the numbers that we had before. On the other hand, the rows (by definition) add up to right-to-left minima statistics: the ith row is the number of right-to-left minima in the permutation occupying the first $N - i + 1$ positions of the array. If we use the same correspondence as before and generate the permutations of N elements by adding a new element at the end of permutations of $N - 1$ elements, then it is plain that the average increase in B is just one less than the number of right-to-left minima in a random permutation of N elements. (The increase simply comes from the 1s in the first row of the table.) This gives a direct solution for the average:

$$B_N = B_{N-1} + H_N - 1$$

$$= \sum_{1 \leq k \leq N} H_k - N$$

$$B_N = (N + 1)H_N - 2N$$

(The sum of the harmonic numbers can be evaluated by substituting in the definition $H_k = \sum_{1 \leq j \leq k} 1/j$, then switching the order of summation: we will see another method at the end of this chapter.) Note carefully that the amount contributed by the Nth element is *not* independent of the arrangement of the previous elements, so we cannot get a "simple" direct recurrence of the generating function by using this simple way to compute the average. Or, put another way, there does not seem to be a simple way to organize the permutations to derive a recurrence in the probabilities that $B = k$. However, Knuth (1971) found enough structure in this problem to derive the variance. Knuth's derivation is rather complicated, but we can develop a simple argument using combinatorial generating functions based on his way of splitting the problem.

Again, we write down the generating function in terms of the combinatorial structure being analyzed, in this case permutations. We have

$$B(w, z) = \sum_{\text{all perms } P} w^{|P|} \frac{z^{v(P)}}{|P|!}$$

where $v(P)$ is the value of B when the algorithm is run on P. As above, our goal is to derive a functional equation using this combinatorial definition, because that can be used to find the first two derivatives with respect to z of $B(w, z)$, evaluated at 1, from which the mean and variance can be computed. Given two permutations P_L, P_R, we can create

$$\binom{|P_L| + |P_R|}{|P_L|}$$

permutations of size $|P_L| + |P_R| + 1$ all of which have the same value of B by (i) concatenating them with a 1 in between; and (ii) making the values distinct by choosing $|P_L|$ values from $2, 3, \ldots, |P_L| + |P_R| + 1$ in all possible ways for assignment to P_L. All of the permutations formed in this way have the same value of B when the *in-situ* permutation algorithm is run on them: $v(P_L) + v(P_R) + |P_R|$. This follows from examining the table above: the 1 has all 1s below it; the elements to the left of these 1s are all 0; and this block divides the table into two independent parts, one for the left permutation, one for the right permutation. Since every permutation of size $|P_L| + |P_R| + 1$ is formed exactly once by this construction, we have the following double sum representation for $B(w,z)$:

$$\sum_{\text{all perms } P_L} \sum_{\text{all perms } P_R} \binom{|P_L| + |P_R|}{|P_L|} \frac{w^{|P_L|+|P_R|+1}z^{v(P_L)+v(P_R)+|P_R|}}{(|P_L| + |P_R| + 1)!} + 1$$

As in the analysis of the internal path length of binary search trees, this reduces, after differentiating by w and rearranging terms, to the simple functional equation:

$$B_w(w, z) = B(w, z)B(wz, z)$$

Then, differentiating by z and evaluating at 1 gives differential equations in the generating functions for the mean and variance that can be solved as in the analysis of binary search trees. (This part of the calculation, involving tricky manipulations with partial derivatives, is by no means simple, but it is so automatic as to be suitable for a computerized symbolic manipulation system such as MACSYMA (Mathlab Group, 1977).

Many interesting combinatorial problems derive from the "cycle" structure of permutations, some of which have direct relevance to various algorithms. For example, the above algorithm, might be slightly improved by a special test for singleton cycles. This could be easily analyzed, because the average number of cycles of size m for any particular m can be derived. On the other hand, questions such as "what is the length of the largest cycle, on the average?" lead to very difficult combinatorial problems (Shepp and Lloyd, 1966).

9.3 SELECTION SORT

A similar quantity arises in selection sort, a method of sorting by successively "finding the minimum":

```
for j := 1 to N do
  begin
  v := ∞; k := j;
  for i := j to N do
    if A[i] < v then begin v := A[i]; k := i end;
  t := A[j]; A[j] := A[k]; A[k] := t;
  end;
```

The operation of this method on our sample file is diagrammed below: the elements in bold face are the left-to-right minima encountered.

3	8	4	**2**	5	9	6	**1**	7
1	**8**	**4**	**2**	5	9	6	**3**	7
1	2	**4**	8	5	9	6	**3**	7
1	2	3	**8**	5	9	6	**4**	7
1	2	3	4	**5**	9	6	8	7
1	2	3	4	5	**9**	**6**	8	7
1	2	3	4	5	6	**9**	**8**	**7**
1	2	3	4	5	6	7	**8**	9
1	2	3	4	5	6	7	8	**9**

The only "variable" in the running time of this program is B, the total number of left-to-right minima encountered during the life of the sort. (This is not the "leading" term in the running time, since the **if** statement is executed $(N + 1/2)$ times.) As above, we can define a correspondence between permutations as follows: given a permutation of $N - 1$ elements, create N permuations of N elements by incrementing each element and then exchanging each element with a prepended 1. For example, 3 1 2 corresponds to:

1	4	2	3
4	1	2	3
2	4	1	3
3	4	2	1

Each of these permutations will result is 1 4 3 2 when the algorithm is iterated once, which is equivalent to 3 1 2 for subsequent iterations. This correspondence immediately implies that

$$B_N = B_{N-1} + H_N = (N + 1)H_N - N$$

But again, we do *not* have independence: for example, if we have a low number of left-to-right minima on one iteration, we can expect a low number on the next. What is worse, each iteration modifies the permutation not yet seen (by exchanging a new element into the position occupied by the current minimum). This "dynamic" aspect seems to make this a much more difficult problem than the *in-situ* permutation problem: the variance of B is not yet known.

9.4 INSERTION SORT

A simple sorting program which can be more successfully analyzed is insertion sort. This method works by "inserting" each element into proper position among those previously considered moving larger elements over one position to make room:

```
A[0] := - ∞;
for i := 2 to N do
    begin
    v := A[i]; j := i−1;
    while A[j] > v do begin A[j + 1] := A[j]; j := j − 1 end;
    A[j + 1] := v;
    end
```

The table below shows the operation of this program on our sample file; elements in bold face are those that are moved.

3	8	4	2	5	9	6	1	7
3	8							
3	4	**8**						
2	**3**	**4**	**8**					
2	3	4	5	**8**				
2	3	4	5	8	9			
2	3	4	5	6	**8**	**9**		
1	**2**	**3**	**4**	**5**	**6**	**8**	**9**	
1	2	3	4	5	5	7	**8**	**9**

The only "variable" in the running time of the program (call it B) is the total number of elements moved: this is the number of elements to the left of each element which are greater than that element, shown for our example in the following table:

$$
\begin{array}{ccccccccc}
3 & 8 & 4 & 2 & 5 & 9 & 6 & 1 & 7 \\
0 & 0 & 1 & 3 & 1 & 0 & 2 & 7 & 2
\end{array}
$$

(Note that this table can be built by considering the permutation as a static object, even though the elements involved may move around before the move which is counted.) This table is a well-known combinatorial object called an *inversion table*, and the total of the numbers in the table is called the number of *inversions* in the permutation. The important properties of inversion tables are that a table uniquely determines the corresponding permutation and that, for a random permutation, the ith entry can take on each value between 0 and $i - 1$ with probability $1/i$, independently of the other entries. This gives a direct generating function derivation for the average number of inversions: the generating function for the number of inversions involving the Nth element is $(1 + z + z^2 + \ldots + z^{N-1})/N$, independent of the arrangement of the previous elements, so that the generating function for the total number of inversions in a random permutation of N elements is given by

$$
B_N(z) = \frac{1 + z + z^2 \ldots + z^{N-1}}{N} B_{N-1}(z)
$$

Again, the mean and variance can be calculated by summing individual means and variances: the mean turns out to be $N(N - 1)/4$, and the variance $N(N - 1)(2N + 5)/72$.

It is an interesting exercise to derive these results using "combinatorial" double generating functions as we have done for several other problems.

9.5   DIGRESSION: DISCRETE SUMS

The calculation of the variance of the number of inversions involves evaluating sums of the form

$$
\sum_{0 \leq k \leq n} k \quad \text{and} \quad \sum_{0 \leq k \leq n} k^2
$$

Since we will be encountering more complicated sums of a similar nature, it is appropriate to consider methods of evaluating such sums at this point. It turns out that a few identities and techniques are sufficient for the evaluation of many sums involving binomial coefficients, harmonic numbers and polynomials.

For polynomials in the index of summation, the fundamental formula on binomial coefficients

$$\sum_{0 \le k < n} \binom{k}{m} = \binom{n}{m+1}$$

should be used. For example

$$\sum_{0 \le k < n} k^2 = \sum_{0 \le k < n} 2\binom{k}{2} + \sum_{0 \le k < n} \binom{k}{1} = 2\binom{n}{3} + \binom{n}{2} = \frac{n(n - \frac{1}{2})(n-1)}{3}$$

In general, a polynomial in k can be expressed as a sum of binomial coefficients, the fundamental formula applied, and the binomial coefficients converted back to polynomials if desired. This identity is akin to a simple integration identity: binomial coefficients are sometimes written in terms of "falling factorial" powers $k^{\underline{m}} = k(k - 1) \ldots (k - m + 1)$, so $m!\binom{k}{m} = k^{\underline{m}}$ and our identity is

$$\sum_{0 \le k < n} k^{\underline{m}} = \frac{n^{\underline{m+1}}}{m+1}$$

which is exactly analogous (for $m \ge 0$) to

$$\int_0^n x^m \, dx = \frac{n^{m+1}}{m+1}$$

This analogy to integration holds elsewhere. For example, there is a "summation-by-parts" formula which can be used to simplify some complicated sums:

$$\sum_{0 \le k < n} (a_{k+1} - a_k) b_k = a_n b_n - a_0 b_0 - \sum_{0 \le k < n} a_{k+1}(b_{k+1} - b_k)$$

This corresponds to the familiar

$$\int_0^n u \, dv = uv \Big|_0^n - \int_0^n v \, du$$

The 'difference" $a_{k+1} - a_k$ in the discrete case corresponds to the "differential" operator dv in the continuous case. For example, summation by parts can be used to sum the harmonic numbers: taking $a_k = k$, $b_k = H_k$, we get

$$\sum_{1 \le k < n} H_k = nH_n - \sum_{1 \le k < n} (k + 1)(H_{k+1} - H_k) = nH_n - n$$

This is the analog to

$$\int_1^n \ln x \, dx = n \ln n - n + 1$$

(In fact, the definition of H_n itself is the analog to $\int_1^n 1/x \, dx = \ln n$.)

Similarly, we can evaluate

$$\sum_{1 \le k < n} \binom{k}{m} H_k$$

in a manner analogous to integrating

$$\int_1^n x^m \ln x \, dx$$

using summation by parts.

The analogy is not a method for evaluating sums (it sometimes breaks down), but it is useful in allowing one to apply intuition about integrals towards sums. One further feature: the analog to e^x is 2^k, since

$$\sum_{0 \le k < n} 2^k = 2^n - 1$$

and thus we can do sums such as $\sum k^2 2^k$, etc.

Later we will see more details on the important class of sums involving two binomial coefficients. Sums with the index of summation appearing in the lower index of a single binomial coefficient are more difficult: although we know that

$$\sum_k \binom{n}{k} = 2^n$$

by the binomial theorem, the partial sum

$$\sum_{0 \le k < m} \binom{n}{k}$$

is very difficult to evaluate. We will see techniques for such an evaluation later.

9.6 TWO-ORDERED PERMUTATIONS

A practical improvement to insertion sort, called Shellsort, reduces the running time well below N^2 by making several passes through the file, each time sorting h independent subfiles (each of size about N/h) of elements spaced by h. This is easily implemented as follows:

```
for h : = h_t, h_{t-1}, ... h_1 do
  for i := h + 1 to N do
    begin
    v := A[i]; j := i - h;
    while j > 0 and A[j] > v do begin A[j + h] := A[j]; j := j - h end;
    A[j + h] := v;
    end;
```

The "increments" $h_t, h_{t-1}, \ldots h_1$ which control the sort must form a decreasing sequence which ends in 1. Considerable effort has gone into finding the best sequence of increments, with few analytic results: although it is a simple extension to insertion sort, Shellsort has proven to be extremely difficult to analyze. This may be appreciated by attempting to analyze the simplest version in which h takes on only two values, 2 and 1.

For example, suppose that the input array is initially

$$3 \quad 5 \quad 10 \quad 6 \quad 1 \quad 9 \quad 14 \quad 2 \quad 15 \quad 12 \quad 11 \quad 8 \quad 16 \quad 7 \quad 4 \quad 13$$

Then, after the first pass, with $h = 2$, the file becomes

$$1 \quad 2 \quad 3 \quad 5 \quad 4 \quad 6 \quad 10 \quad 7 \quad 11 \quad 8 \quad 14 \quad 9 \quad 15 \quad 12 \quad 16 \quad 13$$

Such a permutation, which consists of two interleaved sorted permutations, is called *2-ordered*. Since the next pass of Shellsort, with $h = 1$, is just insertion sort, its running time will vary with the number of inversions. We need to find the average number of inversions in a 2-ordered permutation. This will not only involve some interesting manipulations with finite sums but will also yield some results that will be of use later, since 2-ordered permutations arise in the analysis of several interesting algorithms (for example, they naturally model the input to merging algorithms).

To compute the average number of inversions in a 2-ordered file, we count the total number of inversions appearing in all 2-ordered files, by counting, for each i, the total number of inversions from all 2-ordered files involving the ith element from the odd part of the array. Now, this element can have value $i + j$ for $0 \le j \le N$. Since exactly i of the elements less than $i + j$ appear in the odd part of the array, all the elements in the first j positions of the even part of the array must have values less than $i + j$. From this it is easy to see that the number of inversions to be counted is $i - j - 1$ for $i > j$ and $j - i + 1$ for $j \ge i$, or simply $|i - j - 1|$. Since there are exactly

$$\binom{i + j - 1}{j}\binom{2N - i - j}{N - j}$$

2-ordered files where the ith element in the odd part has the value $i + j$, we are led directly to the formula

$$\binom{2N}{N}A_N = \sum_{1 \le i \le N} \sum_{0 \le j \le N} |i - j - 1|\binom{i + j - 1}{j}\binom{2N - i - j}{N - j}$$

$$= \sum_{0 \le i < N} \sum_{0 \le j \le N} |i - j|\binom{i + j}{i}\binom{2N - i - j - 1}{N - j}$$

for the average number of inversions in a 2-ordered permutation of length $2N$. (Knuth, 1973b, gives a graphical proof of this formula based on a correspondence between 2-ordered permutations and paths in an N by N lattice.)

The general strategy in evaluating this sum is to use symmetries to eliminate the absolute value and to get an inner sum involving only the two binomial coefficients, then evaluate that sum and simplify. In this case, the resulting inner sum is rather difficult to evaluate although it can be done with elementary techniques.

First, to "use symmetries" means to split the inner sum into two parts:

$$\sum_{0 \le i < N} \sum_{0 \le j \le N} (\;) = \sum_{0 \le i < N} \sum_{0 \le j \le i} (\;) + \sum_{0 \le i < N} \sum_{i < j \le N} (\;)$$

then change i to $N - 1 - i$ and j to $N - j$ in the second sum and recombine to get

$$\binom{2N}{N} A_N = \sum_{0 \le i < N} \sum_{0 \le j \le i} (2(i - j) + 1) \binom{i + j}{i} \binom{2N - i - j - 1}{N - j}$$

Now, change j to $i - j$ in the inner sum, interchange the order of summation, then change i to $i + j$ to get

$$\binom{2N}{N} A_N = \sum_{0 \le j < N} (2j + 1) \sum_i \binom{2i + j}{i} \binom{2N - 2i - j - 1}{N - i - j - 1}$$

Below it is shown that

$$\sum_i \binom{2i + j}{i} \binom{2N - 2i - j - 1}{N - i - j - 1} = \sum_{0 \le k < N - j} \binom{2N}{k}$$

$$= \sum_{j < k \le N} \binom{2N}{N - k}$$

which implies, after interchanging the order of summation, that

$$\binom{2N}{N} A_N = \sum_{k \ge 1} \binom{2N}{N - k} \sum_{0 \le j < k} (2j + 1)$$

Note that the only property of $|i - j|$ that we have made use of in this derivation is that it is constant along diagonals; in particular, the derivation works for any function $f(i, j)$ satisfying

$$f(i, i - j) = f(j, 0)$$

and

$$f(i, i + j) = f(0, j) \quad \text{for} \quad j \ge 0$$

For any weight function with these properties, we have proved the following combinatorial identity which (as we will see later) is useful in the study of 2-ordered permutations:

$$\sum_{0 \le i < N} \sum_{0 \le j \le N} f(i,j) \binom{i+j}{i} \binom{2N-i-j-1}{N-j}$$

$$= \sum_{k \ge 1} \binom{2N}{N-k} \sum_{0 \le j < k} (f(j,0) + f(0,j+1))$$

Returning to the number of inversions in a 2-ordered permutation, we find that the inner sum evaluates to k^2, so that

$$\binom{2N}{N} A_N = \sum_{k \ge 1} \binom{2N}{N-k} k^2$$

The easiest way to evaluate this sum is to "absorb" the k^2 into the binomial coefficient by writing it as $N^2 - (N-k)(N+k)$ so that

$$\binom{2N}{N-k} k^2 = N^2 \binom{2N}{N-k} - 2N(2N-1)\binom{2N-2}{N-k-1}$$

Now we are left with sums on the bottom index of a binomial coefficient which can be evaluated since they are nearly over the whole range:

$$\sum_k \binom{2N}{k} = 2^{2N} = 2 \sum_{k \ge 1} \binom{2N}{N-k} + \binom{2N}{N}$$

The final result, which comes after some calculations from the previous three equations, is

$$\binom{2N}{N} A_N = N \, 4^{N-1}$$

so A_N is approximately equal to $\sqrt{\pi N^3}/4$.

It is somewhat surprising that such a simple result requires such a long and complicated derivation: this result deserves a one-line proof! A shorter proof is available through a combinatorial generating function argument: using Knuth's correspondence to paths in a lattice diagram it is possible to show that the generating function

$$B(w,z) = \sum_{\substack{\text{all 2-ordered} \\ \text{perms } P}} w^{|P|} z^{inv(P)}$$

(where $inv(P)$ is the number of inversions in P) satisfies

$$B_w(w,z)|_{z=1} = \frac{w}{(1-4w)^2}$$

However, this derivation not only involves an indirect argument using the generating function for particular types of paths in the lattice but also some complicated manipulations with derivatives of these generating functions (see Knuth, 1973b, Ex. 5.2.1-15). A direct, simple proof of this result seems to be an elusive goal.

9.7   SUMS INVOLVING TWO BINOMIAL COEFFICIENTS

Many sums involving two binomial coefficients reduce in some way to the fundamental Vandermonde convolution:

$$\sum_k \binom{r}{k}\binom{s}{n-k} = \binom{r+s}{n}$$

This is trivially proved with generating functions from the identity

$$(1+z)^r(1+z)^s = (1+z)^{r+s}$$

since $(1+z)^r$ is the generating function for $\binom{r}{k}$, etc. Many other sums involving two binomial coefficients can be derived in this way: for example, Vandermonde's convolution on the upper index

$$\sum_{0\le k\le r} \binom{r-k}{m}\binom{s+k}{n} = \binom{r+s+1}{m+n+1}$$

comes from the simple identity

$$\frac{1}{(1+z)^r}\frac{1}{(1+z)^s} = \frac{1}{(1+z)^{r+s}}$$

A similar (though much more complicated) argument can be used to remove one binomial coefficient from the inner sum which arose above:

$$\sum_i \binom{2i+j}{i}\binom{2N-2i-j-1}{N-i-j-1} = \sum_{i\ge 0}\binom{2N-1-i}{N-i-j-1}2^i$$

Now Vandermonde's convolution can be used in an unexpected way: since

$$2^i = \sum_k \binom{i}{k},$$

the sum is

$$\sum_k\sum_{i\ge 0}\binom{2N-1-i}{N-i-j-1}\binom{i}{k} = \sum_k\sum_{i\ge 0}\binom{2N-1-i}{N+j}\binom{i}{k}$$

$$= \sum_{k\ge 0}\binom{2N}{N+j+k+1} = \sum_{0\le k<N-j}\binom{2N}{k}$$

This derivation is typical: many more examples and many more techniques are given in Knuth 1973a).

9.8 SHELLSORT

Some aspects of the above analysis can be extended to analyzing the general Shellsort algorithm, but the full treatment of this program remains an unsolved problem. Yao has recently done an analysis of $(h, k, 1)$ Shellsort using techniques similar to those above, but the results and methods become much more complicated. For general Shellsort, the functional form is not even known: some conjecture it to be N^α, for some small constant α; others conjecture $N(\log N)^2$. This problem has an interesting combinatorial quality and direct practical relevance (since Shellsort is the method of choice for medium-size files, especially if space for the program is limited). For many more details on this problem and an example of the extensive use of the techniques of this section see Yao (1980).

10. Elementary asymptotic approximations

The methods and examples that we have studied to this point have been oriented towards deriving *exact* average-case results. Unfortunately, such exact solutions may not be always available, or if available they may be too unwieldy to be of much use. In this section, we examine some methods of deriving approximate solutions to problems or of approximating exact solutions.

We have seen that the analysis of computer algorithms involves tools from discrete mathematics, leading to answers most easily expressed in terms of discrete functions (such as harmonic numbers or binomial coefficients) rather than more familiar functions from analysis (such as logarithms or powers). However, it is generally true that these two types of functions are closely related, and one reason to do asymptotic analysis is to "translate" between them. It is sometimes necessary to do such translations in order to invoke more powerful analytic tools.

Generally, the "size" of a problem to be solved is expressed in terms of one (perhaps a few) parameters, and we are interested in approximations that become more accurate as the parameters become large. By the nature of the mathematics and the problems that we are solving, it is also often true that our answers, if they are to be expressed in terms of a single parameter N, will be convergent power series in N and $\log N$. Therefore, our general approach will be to convert quantities to truncated versions of such power series, then manipulate them in well-defined ways. When results cannot be so expressed, then we will need much more powerful tools, as we will see in the next chapter.

168

10.1 *O*-NOTATION

The standard way to write down precise asymptotic approximations is the so-called O-notation. For example, it is shown below that

$$H_N = \ln N + \gamma + \frac{1}{2N} + O\!\left(\frac{1}{N^2}\right)$$

(Here γ is a constant with approximate value $0.57721\ldots$) The quantity represented by the $O(1/N^2)$ term (the error in the approximation) is less in absolute value than some constant divided by N^2 for large enough N. A precise definition of the O-notation may be found in Knuth (1973b, 1976). It is easy to check many elementary properties of the O-notation which facilitate simple calculations involving *asymptotic formulas* such as the one above. For example, consider the formula derived in Chapter 8 for the average internal path length of binary search trees:

$$(N + 1)C'_N = 2(N + 1)H_N - 2N$$

Since $N\,O(1/N^2) = O(1/N)$ and $1/N + O(1/N)$ can be replaced by $O(1/N)$, etc., we have

$$(N + 1)C'_N = 2N \ln N + N(2\gamma - 2) + 1 + O\!\left(\frac{1}{N}\right)$$

No matter what the value of the constant involved in the O-notation (in principle it could be large, in practice it is small), the error in this approximation becomes small as N grows.

The O-notation is useful because it can allow suppression of unimportant details without loss of mathematical rigor or precise results. Note that we could determine C_N to within $O(1/N^2)$ with only a few more calculations. If a more accurate answer is desired, one can be obtained, but most of the detailed calculations are suppressed otherwise. We shall be most interested in methods that allow us to keep this "potential accuracy", producing answers which could be calculated to arbitrarily fine precision if desired.

10.2 BATCHER'S ODD–EVEN MERGE

An example of a problem for which the exact answer is surely very complicated is determining the average number of exchanges required by Batcher's "odd–even" merging method. In this section, we examine the derivation of an approximate answer for this problem that illustrates many basic techniques used for estimating combinatorial sums. In the next section we will study the more powerful methods that are needed to make the answer precise. Details on this derivation may be found in Sedgewick (1978b).

Batcher's method is a way to sort a 2-ordered permutation on $2N$ elements (which is equivalent to doing an N by N merge) as follows:

for $j := 1$ **to** N **do** *compexch(A[2j − 1], A[2j]);*
for $k := lg(N)$ **downto** 1 **do**
        for $j := 1$ **to** $N − 2^{k-1}$ **do** *compexch(A[2j],A[2j + 2^k − 1]);*

Here *compexch* is a procedure which compares its two arguments and exchanges them, if necessary, to make the first smaller. The operation of this method on a sample file is diagrammed below:

1	2	3	5	4	6	10	7	11	8	14	9	15	12	16	13
1	2	3	5	4	6	7	10	8	11	9	14	12	15	13	16
1	2	3	5	4	6	7	10	8	11	9	14	12	15	13	16
1	2	3	5	4	6	7	9	8	11	10	13	12	15	14	16
1	2	3	4	5	6	7	8	9	10	11	12	13	14	15	16

This algorithm seems mysterious at first glance, but actually its operation can be quite easily understood in several ways (see Knuth 1973b; Sedgewick 1978b, 1982a for details). It can be shown that the first stage ensures that the ith element in the odd part of the array has value $i + j$ with $j \le i$, then the subsequent stages ensure that $(i − j) < 2^k$, until reaching the state where the ith element in the odd part of the array contains $i + j$ with $j = i − 1$ for each i, at which point the array is sorted.

Note that the sequence of compare–exchange operations is predetermined (independent of the data) and that all the operations in each stage can be done in parallel, so the method is appropriate for implementation in hardware.

The only "unknown" quantity in the above program is the number of times exchanges are actually done: we will denote the average value of this quantity for a file of size $2N$ by B_N. In Sedgewick (1978b) it is shown that our combinatorial identity for 2-ordered permutations from the previous section holds, with a "weight function" $f(k)$ that turns out to be equal to the number of 1s in the Gray code representation of k (see Ramshaw and Flajolet, 1980). Thus, by that derivation, the average number of exchanges is given by

$$B_N = \sum_{k \ge 1} \frac{\binom{2N}{N-k}}{\binom{2N}{N}} F(k) \quad \text{where} \quad F(k) = 2 \sum_{0 \le j < k} f(j) + k$$

This sum is essentially a sum over the lower index of a binomial coefficient which, as we saw when calculating the number of inversions in a 2-ordered permuation, can be easy to evaluate if $F(k)$ is a simple polynomial in k. However, for other $F(k)$, such sums are difficult to evaluate exactly, and we must resort to asymptotic techniques. To approximate B_N, our method will be to estimate the summand in terms of classical functions, then approximate the sum with an integral, and integrate. We will describe the method in general terms, for it is applicable to a variety of similar sums involving factorials, powers, and other functions, and reasonably well behaved $F(k)$.

A precise way to estimate sums with integrals was given by Euler:

$$\sum_{1 \le k < n} f(k) = \int_1^n f(x)\,dx + \sum_{1 \le k \le m} \frac{B_k}{k!} f^{(k-1)}(x)\big|_1^n + R_m$$

Here B_k are the Bernoulli numbers ($B_0 = 1, B_1 = -1/2, B_2 = 1/6, B_3 = 0$), n is the number of terms desired in the asymptotic estimate, and R_m is a term smaller in absolute value than the last term estimated. (The Bernoulli numbers grow to be quite large, so this formula is typically useful only for small m.) Of course, the function f must be differentiable enough times to make the formula valid.

For example, taking $f(k) = 1/k$ leads to an asymptotic expansion for the harmonic numbers:

$$H_N = \ln N + \gamma + \frac{1}{2N} - \frac{1}{12N^2} + O\left(\frac{1}{N^4}\right)$$

and taking $f(k) = \ln k$ leads to an expansion for $N!$:

$$\ln N! = (N + \tfrac{1}{2})\ln N - N + \ln \sqrt{2\pi} + \frac{1}{12N} + O\left(\frac{1}{N^3}\right)$$

These expansions require calculation of "Euler's constant" $\gamma = 0.57721\ldots$ and "Stirling's constant" $\ln \sqrt{2\pi}$: see Knuth (1973a) for many more details about these expansions and Euler's formula.

To use Euler's formula to approximate B_N, we need to approximate the summand with functions that we could hope to integrate. The first step, which is unnecessary for this problem but very useful for many others, is to "normalize" the sum by centering it so that the largest term occurs for $k = 0$, then dividing by that term. This is useful because not all the terms of the sum contribute significantly to the result. (In fact, as we will see, for this example most do not.)

Next, we need to approximate the summand. For this, we need only Stirling's approximation

$$\ln N! = (N + \tfrac{1}{2})\ln N - N + \ln \sqrt{2\pi} + O\left(\frac{1}{N}\right)$$

Applying this to the binomial coefficients in our summand, we find that

$$\frac{N!N!}{(N+k)!(N-k)!} = \exp\{2 \ln N! - \ln(N+k)! - \ln(N-k)!\}$$

$$= \exp\Big\{(2N+1)\ln N - (N+\tfrac{1}{2})(\ln(N+k)+\ln(N-k))$$

$$- \ln \sqrt{2\pi} - k(\ln(N+k) - \ln(N-k))$$

$$+ O\Big(\frac{1}{N}\Big) + O\Big(\frac{1}{N+k}\Big) + O\Big(\frac{1}{N-k}\Big)\Big\}$$

Note carefully that this approximation is meaningless for $|k|$ close to N; fortunately, we will not need it for k in that range.

The ln function is easily approximated by turning its Taylor series expansions into an asymptotic formula:

$$\ln(1-x) = -x - \frac{x^2}{2} - \frac{x^3}{3} + O(x^4), \quad \text{for} \quad |x| < 1$$

Approximation by Taylor series works for many other functions, for example

$$e^x = 1 + x + \frac{x^2}{2} + \frac{x^3}{6} + O(x^4)$$

Since $\ln(N+k) = \ln N + \ln\Big(1 + \dfrac{k}{N}\Big)$ and similarly for $\ln(N-k)$, we have

$$\ln(N+k) + \ln(N-k) = 2 \ln N - \frac{k^2}{N^2} + O\Big(\frac{k^4}{N^4}\Big)$$

$$\ln(N+k) - \ln(N-k) = \frac{2k}{N} + O\Big(\frac{k^3}{N^3}\Big)$$

Substituting,

$$\frac{N!N!}{(N+k)!(N-k)!} = \exp\Big\{-(N+\tfrac{1}{2})\Big(-\frac{k^2}{N^2} + O\Big(\frac{k^4}{N^4}\Big)\Big) - k\Big(\frac{2k}{N} + O\Big(\frac{k^3}{N^3}\Big)\Big)$$

$$- \ln \sqrt{2\pi} + O\Big(\frac{1}{N}\Big) + O\Big(\frac{1}{N+k}\Big) + O\Big(\frac{1}{N-k}\Big)\Big\}$$

Now we have even more O terms to worry about: we need to restrict k to be less than some value to make all the approximations useful. On the other hand, we cannot restrict k too much, since we will need to deal with the terms for large k in some other way. We will partially postpone this problem as follows. Clearly, since k is our index of summation, we want as few occurrences of k as possible (preferably just one) in our final approximation. The "largest" term involving k is $-k^2/N$; next come $O(k^2/N^2)$ and $O(k^4/N^3)$. These become $O(1/N)$ for $|k| < \sqrt{N}$, so $\sqrt{N}$ is clearly a "critical value" beyond which more values of k may be needed. Recognizing that we may later need to allow k to be larger, we will postpone the exact decision by restricting $|k|$ to be $< \sqrt{Nt(N)}$ for some small function $t(N)$, in which case we have

$$\frac{N!\,N!}{(N+k)!\,(N-k)!} = \exp\left\{\frac{-k^2}{N} + O\left(\frac{t(N)}{N}\right)\right\} = e^{-k^2/N}\left(1 + O\left(\frac{t(N)}{N}\right)\right)$$

(This last equation follows from $e^{O(x)} = 1 + O(x)$, a fact the reader might wish to check.)

Applying the approximation to our sum, we have

$$B_N = \sum_{1 \le |k| \le \sqrt{Nt(N)}} e^{-k^2/N}\left(1 + O\left(\frac{t(N)}{N}\right)\right)F(k)$$

$$+ \sum_{|k| > \sqrt{Nt(N)}} \frac{N!\,N!}{(N+k)!\,(N-k)!}F(k)$$

But we can use the approximation calculated above to bound the second sum, since the terms are decreasing. For $|k| > \sqrt{Nt(N)}$, we know that

$$\frac{N!\,N!}{(N+k)!\,(N-k)!} < \frac{N!\,N!}{(N+\sqrt{Nt(N)})!\,(N-\sqrt{Nt(N)})!}$$

$$= \exp\left\{-t(N) + O\left(\frac{t(N)}{N}\right)\right\}$$

In fact, exactly the same bound holds for $\exp(-k^2/N)$, for $|k| > \sqrt{Nt(N)}$, so we can write

$$B_N = \sum_{k \ge 1} e^{-k^2/N}\left(1 + O\left(\frac{t(N)}{N}\right)\right)F(k) + O\left(\sum_{|k| > \sqrt{Nt(N)}} e^{-t(N)}F(k)\right)$$

Now, if $F(k)$ does not get too large, then $t(N)$ can be chosen large enough to make the second sum small. For example, if $F(k)$ is $O(N^m)$ for some constant m, then $t(N)$ can be chosen to be $(m+2)\ln N$ to give

$$B_N = \sum_{k \ge 1} e^{-k^2/N}F(k)\left(1 + O\left(\frac{\log N}{N}\right)\right) + O\left(\frac{1}{N}\right)$$

As mentioned above, these calculations could, in principle, be carried out to better asymptotic accuracy, but there are complications involved. One problem, as the reader certainly must have noticed, is that it is difficult to predict *a priori* how far to carry out the asymptotic series at various steps in the process: the penalty for taking too few terms is an answer with less accuracy than might be expected; the penalty for taking too many terms is excessive needless calculation.

Now that we have estimated our summand in terms of the exponential function, we can apply Euler's summation formula to estimate it with an integral. In fact, the exponential is so well behaved that this introduces no further error, and we have

$$B_N = \int_1^\infty e^{-x^2/N} F(x) \, dx \left(1 + O\left(\frac{\log N}{N} \right) \right) + O\left(\frac{1}{N} \right)$$

as long as $F(x)$ is well behaved. For example, if $F(x) = 1$, then the integral is the well known normal distribution function (Abramouitz and Stegun 1972), with value $\sqrt{\pi N}$, confirming our earlier estimate for the Catalan numbers. Similarly, for $F(x) = x^2$, the integral is easily evaluated to get an asymptotic formula for the average number of inversions in a 2-ordered file.

For Batcher's merge, it is easy to prove by induction that $F(x) = x \log x + O(x)$. Combined with the previous result, this implies that

$$B_N = \int_1^\infty e^{-x^2/N} x \lg x \, dx + O(N)$$

The substitution $t = x^2/N$ transforms the integral to another well known integral, the "exponential integral function", with the result

$$B_N = \tfrac{1}{4} N \lg N + O(N)$$

Unfortunately, it is not easy to get a better approximation to $F(x)$, and more powerful tools are needed to get a more accurate estimate for B_N, as described in the next section.

However, the general method of approximating a sum by first approximating the summand in terms of simpler functions (using only the large terms), then using Euler's formula to approximate with an integral, is quite often used in asymptotic analysis. The particular example that we have used is a familiar one (corresponding to the normal distribution being the limiting case of the binomial distribution), but the same techniques apply to a wide range of sums. For example, Knuth (1973b) describes how to use the method to evaluate

$$\sum_{k > 0} \frac{N!}{(N - 2k)! \, 2^k k!}$$

which counts the number of permutations consisting of cycles all of length 1 or 2.

10.3 HASHING WITH LINEAR PROBING

For another concrete example of the use of asymptotic methods, we will consider a fundamental strategy for searching: an alternative method to trees called *hashing*.

The idea behind hashing is to try to address directly a set of N keys within a table of size M by using a *hash function* h which maps keys (which have a very large number of possible values) to table addresses (0 to $M - 1$). (A typical way to do this is to use a prime M, then convert keys to large numbers in some natural way and use that number modulo M for the hash value. More sophisticated schemes have been devised.) For example, our sample set of ten keys might have the following hash values:

un	deux	trois	quatre	cinq	six	sept	huit	neuf	dix
4	5	4	6	8	8	5	6	5	8

No matter how good the hash function, some keys will have the same hash values, and a *collision resolution* strategy is needed to decide how to deal with such conflicts. Perhaps the simplest such strategy is *linear probing*: if, when inserting a key into the table, the addressed position (given by the has value) is occupied, then simply examine the previous position. If that is also occupied, examine the one before that, continuing until an empty position is found (if the beginning of the table is reached, simply "cycle" to the end). For the keys above, the table is filled as follows:

0	1	2	3	4	5	6	7	8	9
				un					
				un	deux				
			trois	**un**	deux				
			trois	un	deux	quatre			
			trois	un	deux	quatre		cinq	
			trois	un	deux	quatre	six	**cinq**	
		sept	**trois**	**un**	**deux**	quatre	six	cinq	
	huit	**sept**	**trois**	**un**	**deux**	**quatre**	six	cinq	
neuf	**huit**	**sept**	**trois**	un	**deux**	quatre	six	cinq	
neuf	**huit**	**sept**	**trois**	**un**	**deux**	**quatre**	**six**	**cinq**	dix

Collisions (occupied table positions examined) are printed in bold face in this table. This example (especially the last insertion) shows the algorithm at its worst: it performs badly for a nearly full table, but reasonably for a table with plenty of empty space. As the table fills up, the keys tend to "cluster" together, producing long chains which must be searched to find empty space.

An easy way to avoid clustering is to look not at the previous but at the tth previous position each time a full table entry is found, where t is computed by a second hash function. This method is called double hashing. Other hashing methods use dynamic storage allocation, for example keeping all the keys with the same hash value on a simple linked list.

Linear probing is a fundamental searching method, and an analytic explanation of the clustering phenomenon is clearly of interest. The algorithm was first analyzed by Knuth, who states that this derivation had a strong influence on the structure of his books. His books certainly have had a strong influence on the structure of research in the mathematical analysis of algorithms, and this derivation is a prototype example showing how a simple algorithm can lead to nontrivial and interesting mathematical problems.

Knuth's derivation divides into two parts, a combinatorial argument leading to an exact answer and an asymptotic estimation of that answer. Since the combinatorial argument is long, largely specific to this problem, and similar to things we have done in previous chapters, it will be only summarized here. Some of the asymptotic estimation is similar to the above, so it also will be summarized here. Full details are found in Knuth (1973a, 1973b).

We are interested in knowing the average number of table entries examined for a successful search, after N keys have been inserted into an initially empty table of size M. Knuth shows that this quantity is given by

$$C_{NN} = \frac{1}{N} \sum_{0 \le i < N} \sum_{0 \le k \le i} \sum_{k \le j \le i} \frac{k+1}{M^i} \binom{i}{j}\left(\frac{1}{j+1}\right)$$
$$\times (j+1)^j \left(\frac{M-i-1}{M-j-1}\right)(M-j-1)^{i-j}$$

This formula looks formidable, but the inner sum is actually quite similar to the sum in *Abel's binomial theorem*, which says that

$$(x+y)^i = \sum_j \binom{i}{j} x(x-jz)^{j-1}(y+jz)^{i-j}$$

or, with appropriate substitutions,

$$m^i = \sum_j \binom{i}{j}(j+1)^{j-1}(m-j-1)^{i-j}$$

Abel's theorem turns out to be a powerful enough tool to evaluate this sum, although a substantial amount of computation is involved: the details may be found in Knuth (1973b). The eventual result is

$$2C_{MN} - 1 = \sum_{0 \le i \le N} \frac{N!}{M^i(N - i)!}$$

(This formula seems a perfect candidate for derivation using triple "combinatorial" generating functions, but such a derivation has not yet been worked out.) This is an exact answer, but it is not in a particularly useful form, so that we would like to do an asymptotic estimation.

Practically speaking, it seems clear that the table should not be allowed to get nearly full: if we define $N/M = \alpha$, we certainly want $\alpha \ll 1$. If α is not too close to 1, the sum is not difficult to estimate using the same techniques as above:

$$2C_{MN} - 1 = \sum_{0 \le i \le N} \frac{N!}{M^i(N - i)!} = \sum_{0 \le i \le N} \left(\frac{N}{M}\right)^i \frac{N!}{M^i(N - i)!}$$

Splitting the sum into two parts, etc., exactly as above, we can use the fact that terms in this sum begin to get negligibly small after $i > \sqrt{N}$ to prove that

$$2C_{MN} - 1 = \sum_{i \ge 0} \left(\frac{N}{M}\right)^i \left(1 + O\left(\frac{i^2}{N}\right)\right) = \frac{1}{1 - \alpha} + O\left(\frac{1}{N(1 - \alpha)^3}\right)$$

Thus, for fixed α, as N (and M) get large, the average successful search cost for linear probing is about $1/(1 - \alpha)$.

10.4 LINEAR PROBING IN A FULL TABLE

The constants implicit in the O-notation in the equation above are independent of N and α, but the equation becomes meaningless as α gets close to 1. For simplicity, consider the case $N = M$, or

$$2C_{NN} - 1 = \sum_{0 \le i \le N} \frac{N!}{N^i(N - i)!}$$

In this case, it will be convenient to rewrite the sum as

$$2C_{NN} - 1 = \sum_{i} \binom{N}{i} \frac{i!}{N^i}$$

The methods above could be used, even in this case, to eventually estimate the sum with an integral, but it will be instructive to consider a somewhat more direct method which makes use of the Γ-function, defined by

$$\Gamma(i + 1) = \int_0^\infty e^{-t}t^i \, dt$$

This is a generalization of the factorial: it is not difficult to prove that $\Gamma(i + 1) = i\Gamma(i)$ and $\Gamma(i + 1) = i!$, using integration by parts. It is a generalization because it is defined even for noninteger i. This formula becomes directly useful in our problem if we take $t = uN$ so that

$$\frac{\Gamma(i + 1)}{N^{i+1}} = \int_0^\infty e^{-Nu}u^i \, du$$

This is equal to $i\,!/N^{i+1}$ for integer i, so it can be substituted directly into our sum to yield

$$2C_{NN} - 1 = \sum_i \binom{N}{i}N\int_0^\infty e^{-Nu}u^i \, du$$

Now we can interchange the order of integration and summation to get

$$2C_{NN} - 1 = N\int_0^\infty e^{-Nu}(1 + u)^N \, du$$

and we need only evaluate the integral. Knuth (1973a, Section 1.2.11.3) shows how to get an asymptotic value for the integral by using reversion of power series: if we change variables to $v = u - \ln(1 + u)$, so that $(1 + u)^N = e^{N(u-v)}$ and $du = [1 + (1/u)]dv$, then we can expand the logarithm to get

$$v = \tfrac{1}{2}u^2 - \tfrac{1}{3}u^3 + O(u^4)$$

and solve for u, "bootstrapping" one term at a time, to get

$$u = \sqrt{2v} + \tfrac{2}{3}v + O(v^{3/2})$$

and

$$1 + \frac{1}{u} = \frac{1}{\sqrt{2v}} + \frac{2}{3} + O(\sqrt{v})$$

Knuth gives details and further terms of these expansions, as well as a proof that it is valid to use them within the range of integration under consideration. Thus the integral is

$$2C_{NN} - 1 = N\int_0^\infty e^{-Nv}\left(\frac{1}{\sqrt{2v}} + \frac{2}{3} + O(\sqrt{v})\right) \, dv$$

But now we can use the definition of the Γ function for $i = -\tfrac{1}{2}, 0, \tfrac{1}{2}$, to get the result

$$2C_{NN} - 1 = N\left(\frac{\Gamma(\frac{1}{2})}{\sqrt{2N}} + \frac{2}{3}\frac{\Gamma(1)}{N} + O(N^{-3/2})\right)$$

$$C_{NN} = \sqrt{\pi N/8} + \frac{5}{6} + O(N^{-1/2})$$

($\Gamma(1/2)$ is easily calculated from the normal distribution function.) This result can be extended to show that $C_{MN} = O(\sqrt{N})$ whenever $N > M - \beta$ for fixed β.

The method used here of converting a sum to an integral may seem quite mysterious at first (especially since we used the same formula to evaluate the integral); in the next chapter we see that it is a special case of a powerful general technique which is applicable to many problems.

11. Asymptotics in the complex plane

The analysis of many algorithms cannot be understood without resorting to complex numbers. In the simplest cases, roots of polynomials can be involved; in advanced cases, complex analysis is necessary to derive answers which could not otherwise be stated, much less derived. A substantial body of mathematics has been built up, mostly in other contexts, which applies directly to analytic problems associated with simple algorithms. A survey treatment of the "classical analysis" necessary to solve such problems would be far beyond the scope of these notes; nonmathematicians who are convinced by now of the potential utility of such analyses should first read Knopp (1945). Advanced material on some topics relevant to the analysis of algorithms may be found in Bender (1974) and DeBruijn (1958).

11.1 POLYPHASE MERGING

For our first example, we will consider the problem of sorting a file which is too large to fit in our computer's memory but which does fit on a magnetic tape. To sort the contents of a magnetic tape, we will need some auxiliary tapes; a good algorithm will minimize the amount of movement among tapes.

Several methods have been developed for this problem: they all begin by reading the input tape into memory and writing sorted "runs" of information out onto the auxiliary tapes. (The naive method for doing so would result in runs about the size of the internal memory; the actual method usually used manages to create runs about twice the size of the internal memory.) Then the runs are merged together in successive phases to make longer and longer runs, ultimately producing the sorted file.

Many different algorithms have been developed which work this way. A

180

fundamental method is the *polyphase merge* which works (after the runs have been distributed onto the auxiliary tapes in some "perfect" distribution) by taking one run from each tape, and merging them together to make a longer run on an output tape until one tape becomes empty: at this point the process is repeated, using the emptied tape as the new output tape. For example, suppose that three tapes are used and the first step distributes 34 sorted runs onto Tape 1 and 21 onto Tape 2. Then the polyphase "merge until empty" strategy produces a fully sorted file on Tape 2 as follows:

Tape 1	Tape 2	Tape 3	Total runs
34(1)	21(1)	0	55
13(1)	0	21(2)	34
0	13(3)	8(2)	21
8(5)	5(3)	0	13
3(5)	0	5(8)	8
0	3(13)	2(8)	5
2(21)	1(13)	0	3
1(21)	0	1(34)	2
0	1(55)	0	1

The numbers in the table are the number of runs on the tape; the numbers in parentheses are the sizes of the runs (relative to the size of the initial runs). For example, in the second phase, 13 (initial) runs from Tape 1 are merged with 13 of the 21 runs on Tape 3 (of relative size 2) to make 13 runs (or relative size 3) on Tape 2, leaving Tape 1 empty and 8 runs (of size 2) on Tape 3.

The question immediately arises: how many phases are required to merge together N initial runs? For the three-tape case, this is not a particularly difficult question (many readers may already have recognized the Fibonacci numbers), but if more than three tapes are used, the situation becomes more complicated and it is best to use complex analysis.

The merge pattern is easy to derive by working backwards. For example, the following table shows the pattern for four tapes (the table ignores tape labels and is skewed to make the pattern obvious).

1	1	0	0	0			
4		1	1	1	1		
7			2	2	2	1	
13				4	4	3	2	
25					8	7	6	4
49						15	14	12	8

Each line in the table is obtained from the previous line by removing the leftmost entry, adding it to the others, then appending it to the right. It is easy to verify that "merge until empty", starting with these distribution patterns, will implement a polyphase merge.

Polyphase merging and similar methods lead to very interesting and intricate patterns which have been subjected to intensive study and analysis. For example, the problem of how to add "dummy" runs if the number of runs to be merged is not a perfect total appearing in the table has been studied in detail, with the surprising answer that many more dummy runs should be added than seems necessary (see Knuth, 1973b). Our purpose in discussing polyphase merging is to motivate a particular method of analysis, so we will concentrate exclusively on the simplest problem of determining how many phases are required to merge N runs together.

The analysis starts from the observation that each number in the table (including the totals) is the sum of the previous four along the same diagonal: in general, for m tapes, the total number of runs in an n-phase merge is given by the recurrence

$$t_n = t_{n-1} + t_{n-2} + \ldots + t_{n-m+1} \quad \text{for} \quad n > 0$$

with

$$t_0 = t_{-1} = \ldots = t_{2-m} = 1$$

This recurrence directly yields a closed-form expression for the generating function:

$$T_m(z) = \sum_{n>0} t_n z^n = \frac{(m-1)z + (m-2)z^2 + \ldots + z^{m+1}}{1 - z - z^2 - \ldots - z^{m-1}}$$

For example,

$$T_5(z) = \frac{4z + 3z^2 + 2z^3 + z^4}{1 - z - z^2 - z^3 - z^4}$$

and

$$T_3(z) = \frac{2z + z^2}{1 - z - z^2}$$

The standard technique for finding t_n is to expand these expressions for $T_m(z)$ in power series, then equate coefficients of t_n. This can be done by factoring the denominator and then using partial fractions, as illustrated below for $M = 3$.

$$T_3(z) = \frac{2z + z^2}{(1 - \phi z)(1 - \hat{\phi} z)}$$

$$= \frac{2 + z}{\sqrt{5}} \left(\frac{1}{1 - \phi z} - \frac{1}{1 - \hat{\phi} z} \right)$$

$$= \frac{2 + z}{\sqrt{5}} \left(\sum_{n \geq 0} \phi^n z^n - \sum_{n \geq 0} \hat{\phi}^n z^n \right)$$

where

$$\phi = \frac{1 + \sqrt{5}}{2} \quad \text{and} \quad \hat{\phi} = \frac{1 - \sqrt{5}}{2}$$

Therefore,

$$t_n = \frac{1}{\sqrt{5}} (2\phi^n + \phi^{n-1} - 2\hat{\phi}^n - \hat{\phi}^{n-1}) = \frac{\phi^{n+2}}{\sqrt{5}} + O(\hat{\phi}^n)$$

(Note that $\phi + 1 = \phi^2$, so $2\phi + 1 = \phi^3$, and that $|\hat{\phi}| < 1$, so that the contribution of $\hat{\phi}^n$ is negligible compared to that of ϕ^n. Now, if $t_n = N$, we can take logs to find that

$$\log_\phi N = (n + 1) + \log_\phi \left(\frac{1}{\sqrt{5}} \right) + \log_\phi \left(1 + O \left(\frac{\hat{\phi}}{\phi} \right)^n \right)$$

so that the number of phases required to merge together N runs is

$$\log_\phi N - \tfrac{1}{2} \log_\phi 5 - 1 + O(N^{-\varepsilon})$$

for a positive ε depending on ϕ and $\hat{\phi}$.

This technique generalizes. If the roots of $1 - z - z^2 - \ldots - z^{m-1}$ are $\alpha_1, \alpha_2, \ldots, \alpha_{m-1}$, then we can apply partial fractions to give

$$T_m(z) = h_1(z) \sum_{n \geq 0} \alpha_1^n z^n + h_2(z) \sum_{n \geq 0} \alpha_2^n z^n + \ldots + h_{m-1}(z) \sum_{n \geq 0} \alpha_{m-1}^n z^n$$

where $h_i(z)$ are polynomials coming from the partial fractions calculation. It turns out that the inverse of one of the roots say α_1, always dominates the others in the same way that φ dominates $\hat{\varphi}$, even though some of the roots are complex (see Knuth, 1973b, for a proof). Therefore, as above, we eventually find that the number of phases required to merge N runs using M tapes is

$$\log_{\alpha_1} N - \log_{\alpha_1} h_1 \left(\frac{1}{\alpha_1} \right) + O(N^{-\varepsilon})$$

as above. Now, Knuth shows that

$$\alpha_1 = 2 - \frac{1}{2^M} + O \left(\frac{M}{4^M} \right)$$

as M grows, and that all the other roots are smaller. This gives enough information to calculate the leading term. Unfortunately, the next term is available with this method only through a laborious partial fractions calculation involving all the other roots. However, complex analysis provides a convenient method for performing this calculation.

The idea is to note that, as a function in the complex plane, the generating function $T_m(z)$ has a simple pole at $1/\alpha_1$: put another way, $(1 - \alpha_1 z)T_m(z)$ converges for $|z| < R$ with $R > 1/|\alpha_1|$. Therefore,

$$T_m(z) = \frac{H}{1 - \alpha_1 z} + \sum_{n>0} r_n z^n \quad \text{with} \quad |r_n| < R^{-n}$$

As above, this leads to the expression

$$\log_{\alpha_1} N + \log_{\alpha_1} H + O(N^{-\varepsilon})$$

for the number of phases required to merge N runs. The constant H is the "residue" of $T_m(z)$ at $1/\alpha_1$; it can be calculated using l'Hospital's rule. If we denote the numerator of $T_m(z)$ by $p_m(z)$ and the denominator by $q_m(z)$, we have

$$H = \lim_{z \to 1/\alpha_1} (1 - \alpha_1 z)T_m(z) = \lim_{z \to 1/\alpha_1} \frac{(1 - \alpha_1 z)p_m(z)}{q_m(z)} = \frac{\alpha_1 p_m(1/\alpha_1)}{q'_m(1/\alpha_1)} .$$

This calculation is a simple example of a general technique that we will examine more closely below: a function in the complex plane is approximated by another that performs similarly near its singularities. Our next example shows a more complicated application of this method.

11.2 COUNTING ORDERED TREES

Suppose that we wish to count the number of ordered trees with N internal nodes. We know from the direct correspondence between binary trees and ordered trees that the answer is given by the Catalan numbers. This result can also be derived with a direct generating function argument. If $G(z) = \sum_{N \geq 0} g_N z^N$ is the generating function for the number of ordered trees with N internal nodes, then $G(z)^2$ is the generating function for internal nodes in two ordered trees, $G(z)^3$ for three ordered trees, etc. Since every ordered tree consists of a root with one son (which is an ordered tree) or with two sons (which are both ordered trees) or with three sons, etc., we are immediately led to the recurrence

$$G(z) = z(1 + G(z) + G(z)^2 + G(z)^3 + \ldots +) = \frac{z}{1 - G(z)} .$$

This gives a quadratic equation with the solution

$$G(z) = \tfrac{1}{2}(1 \pm \sqrt{1 - 4z})$$

This is very similar to the generating function for binary trees: as before, we choose the root which yields the right small values; then expand using the binomial theorem; then approximate using Stirling's formula to get the result

$$g_{N+1} = \frac{4^N}{N\sqrt{\pi N}}\left(1 + O\!\left(\frac{1}{N}\right)\right)$$

But suppose we want to count the number of ordered trees with N *external* nodes. This is not a well-posed problem because, for example, there are an infinite number of ordered trees with only one external node (any number of unary nodes strung together). This anomaly is fixed by disallowing one-way branching: how many ordered trees are there with no unary nodes and N external nodes? Arguing as above, we immediately get the generating function recurrence:

$$G(z) = z + G(z)^2 + G(z)^3 + \ldots = z + \frac{G(z)^2}{1 - G(z)}$$

This leads to a different quadratic equation with the solution of interest

$$G(z) = \tfrac{1}{4}(1 + z - \sqrt{1 - 6z + z^2})$$

But now we cannot get anywhere by applying the binomial theorem, since it would lead back to a convolution sum to find g_N. To simplify notation, we will consider the asymptotics of $\sqrt{(1 - w)(1 - \alpha w)}$ for $\alpha < 1$: to find g_N we will need to take $w = (3 + 2\sqrt{2})z$ and $\alpha = (3 - 2\sqrt{2})/(3 + 2\sqrt{2})$.

Now, $\sqrt{(1 - w)(1 - \alpha w)}$ has an *algebraic* singularity at $w = 1$: it is not a pole, so we will not be able to approximate the function accurately with negative powers of $(1 - w)$, as above. Nevertheless, we do expect the function to behave something like $\sqrt{(1 - w)(1 - \alpha)}$ at $w = 1$, and this suggests that we compute

$$\sqrt{(1 - w)(1 - \alpha w)} = \sqrt{(1 - w)(1 - \alpha)} + \sqrt{1 - w}\left(\sqrt{(1 - \alpha w)} - \sqrt{1 - \alpha}\right)$$

$$= \sqrt{(1 - w)(1 - \alpha)} + \frac{\alpha(1 - w)^{3/2}}{\sqrt{1 - \alpha w} + \sqrt{1 - \alpha}}$$

The function $(\sqrt{1 - \alpha w} + \sqrt{1 - \alpha})^{-1}$ converges for $w > 1$; therefore, as we have argued before, it is the generating function for a sequence which is $O(r^{-N})$ for some $r > 1$. By the binomial theorem $(1 - w)^{3/2}$ is the generating function for

$$(-1)^N \binom{3/2}{N} = O(N^{-5/2})$$

It remains only to convolve these two sequences: the coefficient of w^N in this convolution is

$$\sum_{0 \le k \le N} O\left(\frac{k^{-5/2}}{r^{N-k}}\right) = \sum_{0 \le k \le N/2} O\left(\frac{k^{-5/2}}{r^{N-k}}\right) + \sum_{N/2 \le k \le N} O\left(\frac{k^{-5/2}}{r^{N-k}}\right)$$

$$= O(r^{-N/2}) + O(N^{-5/2}) \sum_{N/2 \le k \le N} r^{k-N}$$

$$= O(N^{-5/2})$$

This completes the approximation:

$$\sqrt{(1-w)(1-\alpha w)} = \sum_{N \ge 0} \left(\sqrt{1-\alpha} \binom{\frac{1}{2}}{N}(-1)^N + O(N^{-5/2})\right) w^N$$

which eventually leads us to conclude that the number of ordered trees with N external nodes and no unary nodes is approximately $((3\sqrt{2} - 4)/16\pi N^3)^{1/2}(3 + 2\sqrt{2})^n$.

11.3   METHOD OF DARBOUX

Again we were able to estimate a function by subtracting a function which behaves similarly near a singularity, then estimating the error. In this case, the error was not exponentially small (as it was in the case of poles) but merely slightly smaller than the leading term. It turns out that this general method applies to many generating functions. The technique is well known in analysis as the *method of Darboux*: if $G(z) = \sum_{N \ge 0} g_N z^N$ is analytic near α and has only an algebraic singularity of the form $(1 - z/\alpha)^{-w}h(z)$, then

$$g_N = \frac{h(\alpha)N^{w-1}}{\Gamma(w)\alpha^N} + O\left(\frac{N^{w-1}}{r^N}\right).$$

This formula yields directly the three asymptotic results derived above. A more general statement (when more singularities are involved) is given in Bender (1974). This method can give a quick solution to many problems. On the other hand, functions do arise in the analysis of simple algorithms which are too ill-behaved for this theorem to be useful. The general idea of studying the behavior of the function around its singularities is preserved, but much deeper arguments are required; for examples of this see Odlyzko (1979) and Flajolet and Odlyzko (1980).

11.4   MELLIN TRANSFORMS

In the previous section, we were unable to find the coefficient of the linear term in the analysis of Batcher's sort. This problem can be solved using the Mellin integral transform, a technique which has wide applicability in the analysis of algorithms. As we shall see, the answer thus derived has characteristics that make it plain that simpler methods will not work for this problem. Furthermore, several problems from a variety of applications exhibit similar characteristics. This method is outlined in detail in Knuth *et al.* (1969), Knuth (1973b), and Sedgewick (1978b), so we will only sketch it here.

We will pick up the derivation at the point at which we need to evaluate

$$B_N = \sum_{k \geq 1} e^{-k^2/N} F(k)$$

The asymptotic derivation for linear probing in the previous section involved "transforming" a sum involving a factorial (the Γ-function) into an integral involving an exponential. For this problem, we do the opposite: change the sum involving the exponential above into an integral involving a Γ-function.

The specific tool we need to start is the Mellin inversion theorem:
If f is piecewise continuous and

$$F(s) = \int_0^\infty x^{s-1} f(x) \, \mathrm{d}x$$

is absolutely convergent for $c_1 < \operatorname{Re}(s) < c_2$ then

$$f(x) = \frac{1}{2\pi i} \int_{\sigma - i\infty}^{\sigma + i\infty} x^{-s} F(s) \, \mathrm{d}s$$

for $c_1 < \sigma < c_2$. This is a special case of Fourier inversion (see Titschmarsh, 1951) which has applicability in analytic number theory and many other fields.

Taking $f(x) = e^{-x}$ gives $F(s) = \Gamma(s)$ by the definition of the Γ-function (see the previous section), so the inversion theorem says that

$$e^{-x} = \frac{1}{2\pi i} \int_{\sigma - i\infty}^{\sigma + i\infty} x^{-s} \Gamma(s) \, \mathrm{d}s \quad \text{for} \quad \sigma > 0.$$

This formula may be proved independently by noting that the value of the integral is very closely approximated by the sum of the residues to the left of the line of integration because the Γ function is very small for arguments at the fringes of the left half plane. (See Knuth, 1973b for a detailed proof of this: the line integral is approximated by a contour integral which encloses the left half plane in the limit, and the contributions of the other parts of the contour are shown to vanish.) The Γ function has a pole at each nonpositive integer (see Whittaker and Watson, 1927, for properties of the Γ function); the residue at $-j$ is $(-1)^j/j!$. Therefore, the sum of the residues to the left of the line of integration in the integral above is

$$\sum_{j \geq 0} \frac{x^j(-1)^j}{j!} = e^{-x},$$

verifying the relationship between $\Gamma(s)$ and e^x implied by Mellin inversion.

To use Mellin inversion for our sum, we simply "transform" the exponential in our sum,

$$B_N = \sum_{k \geq 1} \frac{1}{2\pi i} \int_{\sigma-i\infty}^{\sigma+i\infty} \left(\frac{k^2}{N}\right)^{-s} \Gamma(s) \, ds \, F(k)$$

then interchange the order of summation and integration to get

$$B_N = \frac{1}{2\pi i} \int_{\sigma-i\infty}^{\sigma+i\infty} \left(\sum_{k \geq 1} \frac{F(k)}{k^{2s}}\right) N^s \Gamma(s) \, ds$$

then "transform" back to get the functional form of B_N. Note that the interchange of integration and summation is allowed only if

$$\sum_{k \geq 1} \frac{F(k)}{k^{2s}}$$

is defined on the path of integration. This depends both on $F(k)$ and σ, and plays an important role in the derivation, as we will see in the examples below. Moreover, we need to be able to study properties of this function in the complex plane. Fortunately, for many examples which arise in the analysis of algorithms, this function can be expressed in terms of classical special functions.

To begin, we will consider an example for which we already know the answer: if we take $F(k) = k^2$, then, as we saw in Chapter 9, B_N is the number of inversions in a 2-ordered permutation. We have

$$\sum_{k \geq 1} \frac{F(k)}{k^{2s}} = \sum_{k \geq 1} \frac{1}{k^{2s-2}} \equiv \zeta(2s - 2)$$

the well known Riemann ζ-function (see, for example, Abramowitz and Stegun, 1972). The sum converges for $s > 3/2$, so we have from above

$$B_N = \frac{1}{2\pi i} \int_{\sigma-i\infty}^{\sigma+i\infty} \zeta(2s - 2) N^s \Gamma(s) \, dx \quad \text{for} \quad \sigma > 3/2$$

Now we can evaluate this integral using a contour integration argument similar to the proof of the Mellin inversion formula described above. Consider

$$\int_R \zeta(2s - 2)N^s\Gamma(s) \, ds$$

where R is a rectangle comprised of two long vertical lines at $\text{Re}(s) = 2$ and $\text{Re}(s) = 1$ and two short horizontal lines connecting them, say at $\text{Im}(s) = \pm M$. The Γ-function becomes exponentially small in the complex part of its argument (see Whittaker and Watson, 1927, for specific bounds), so it is possible to prove, by taking limits as M goes to infinity, that the leading aysmptotic term in the value of the integral is the sum of the residues in the area between $\text{Re}(s) = 1$ and $\text{Re}(s) = 2$. There is only one pole in this strip, at $s = 3/2$, so this leads to a short proof that

$$B_N \approx N^{3/2}\Gamma(3/2) = \sqrt{\pi N^3}/4$$

This might be too powerful a tool for this simple problem, but there are many problems for which Mellin transforms are applicable although more traditional methods fail. Solutions for these problems follow the same general structure as the argument above; significant differences show up in the type of convex integration which must be performed.

For example, in the analysis of radix-exchange sorting, Knuth (1973b) uses the general method above to derive the following integral:

$$B_N = \frac{1}{2\pi i} \int_{-3/2-i\infty}^{-3/2+i\infty} \frac{1}{2^{-1-s} - 1} N^{-s}\Gamma(s) \, ds$$

This integral can be evaluated with residues as above, but a double pole at $s = 1$ is involved, as well as a series of new poles along the line at $\text{Re}(s) = -1$. (For details see Knuth 1973b.) Derivations involving similar functions, with similar complications, may be found in the study of merging networks (Sedgewick, 1978b; Flajolet *et al.*, 1977) and in many other applications. The solutions of these problems take on a complicated form depending on the residue structure of the functions: they are unlikely to be accessible with more elementary techniques. Moreover, as we have seen, Mellin transforms do work properly for some simple problems, so there is some possibility that a general method of solution of a wide class of recurrences based on the Mellin transform could be developed.

12. Probabilistic models

Many algorithms readily admit natural input models. It is quite reasonable to analyze the cost of sorting a randomly ordered file, or searching for a random element, or finding the greatest common divisor of two randomly selected integers. Even for such algorithms, however, it is difficult to be confident that the model used for the analysis reflects reality sufficiently well to allow use of the analysis to predict the performance of algorithms in actual applications. Moreover, sometimes several different models suggest themselves. And for many problems, no "natural" input model is apparent or even likely to exist. In this chapter, we consider some of the difficulties inherent in analyzing algorithms in the face of such uncertainties.

Such difficulties can arise even in more "classical" analyses like those that we have been considering. For example, in our examination of the Quicksort algorithm, we were careful to note that the partitioning process preserves randomness in the sense that it produces random subfiles if used on a random file. Some variants of Quicksort do not have this property, and there are many other algorithms which operate on random inputs in a controlled way that (incidental to the algorithm) destroys the randomness. Some examples of algorithms in this class are balanced tree algorithms for searching (see Guibas and Sedgewick, 1978), and Heapsort (Sedgewick, 1982b). The analysis of such algorithms cannot take advantage of the natural input model, and is thus similar to the analysis of algorithms in the absence of a good input model.

When several different input models are available, how is one to choose among them? Different models may be appropriate for different situations, but the analyst is basically forced to trade off between the ease of analysis and the degree to which the model reflects reality. A description which purports to model the actual expected inputs will be of little us if it is too complicated to admit any analysis, and analytic results on simple artificial models are of limited utility. As we have seen, even very simple algorithms can require quite sophisticated analysis, so it is reasonable to first find a model in which the analysis is tractable, then extend the analysis to more complicated (and more realistic) models.

One approach that is useful when average-case analysis is difficult or impossible is to concentrate on best-case and worst-case performance. This analysis might be easier, less sensitive to assumptions about the input, and still indicative of the performance of the algorithm. For example, it is not difficult to estimate the worst-case behavior of both Heapsort and balanced tree algorithms, and to prove that the average-case performance is within a constant factor of the worst-case. On the other hand, the worst-case performance may be misleading and indicative only of how the algorithm performs for a very few artificially constructed inputs that would never occur in practice. In fact, it is sometimes possible to prove that this situation exists, and take advantage of it to make precise statements about the average behavior of algorithms whose performance characteristics are about the same for virtually all inputs.

Below we will examine some of these issues by studying some algorithms for simple set maintenance. Several variants on a trivial algorithm have been proposed for this problem, and it is not clear which is best. Also, several models have been proposed for the input, and it is not clear which best reflects actual situations in which the algorithms are used.

12.1 UNION-FIND ALGORITHMS

The *union-find* problem involves processing a set of equivalence relations while maintaining an internal data structure which allows testing whether two given items are equivalent. The algorithm is to process two kinds of requests, *union (x, y) for x ≡ y*, and *find(x)*, which is to return the "name" of the equivalence class containing x so that one can test whether x_1 and x_2 are in the same equivalence class by checking whether *find(x_1) = find(x_2)*. For the purposes of analysis we will assume that the elements being processed are the integers 1 to N; in a practical implementation, hashing would be used to convert arbitrary names to this form.

One simple solution to this problem is to maintain an **array** $a[1 . . N]$ containing the name of the equivalence class of each element: to process *union(x, y)*, simply give all elements in x's class the name of y's class. Equivalence class "names" are simply element indices, so we start with $a[i] = i$ to indicate that each element is in its own class. The table below shows how the array changes as the equivalence relations at the left hand side are processed.

	1	2	3	4	5	6	7	8	9	10	11	12	13
	1	2	3	4	5	6	7	8	9	10	11	12	13
4 ≡ 9	1	2	3	9	5	6	7	8	9	10	11	12	13
2 ≡ 11	1	11	3	9	5	6	7	8	9	10	11	12	13
7 ≡ 3	1	11	3	9	5	6	3	8	9	10	11	12	13
6 ≡ 3	1	11	3	9	5	3	3	8	9	10	11	12	13
1 ≡ 5	5	11	3	9	5	3	3	8	9	10	11	12	13
11 ≡ 12	5	12	3	9	5	3	3	8	9	10	12	12	13
3 ≡ 13	5	12	13	9	5	13	13	8	9	10	12	12	13
9 ≡ 10	5	12	13	10	5	13	13	8	10	10	12	12	13
2 ≡ 3	5	13	13	10	5	13	13	8	10	10	13	13	13
4 ≡ 8	5	13	13	8	5	13	13	8	8	8	13	13	13
3 ≡ 5	5	5	5	8	5	5	5	8	8	8	5	5	5
9 ≡ 3	5	5	5	5	5	5	5	5	5	5	5	5	5

This algorithm was called the *quick-find* algorithm by Yao (1976), because the cost of a *find* operation is constant (the time required to access the array). The "variable" in the running time of this algorithm is the number of elements processed during a *union*. In the worst case, this could be large, since it is proportional to the size of one of the sets being merged.

One way to reduce the cost of *union* operations, called the *weighted quick-find* algorithm by Yao, is to change names in the smaller of the two classes to be merged. This requires maintaining an array of weights containing the number of elements in each equivalence class. The table below shows the operation of this version of the algorithm on our example.

	1	2	3	4	5	6	7	8	9	10	11	12	13
	1	2	3	4	5	6	7	8	9	10	11	12	13
4 ≡ 9	1	2	3	9	5	6	7	8	9	10	11	12	13
2 ≡ 11	1	11	3	9	5	6	7	8	9	10	11	12	13
7 ≡ 3	1	11	3	9	5	6	3	8	9	10	11	12	13
6 ≡ 3	1	11	3	9	5	3	3	8	9	10	11	12	13
1 ≡ 5	5	11	3	9	5	3	3	8	9	10	11	12	13
11 ≡ 12	5	11	3	9	5	3	3	8	9	10	11	11	13
3 ≡ 13	5	11	3	9	5	3	3	8	9	10	11	11	3
9 ≡ 10	5	11	3	9	5	3	3	8	9	9	11	11	3
2 ≡ 3	5	3	3	9	5	3	3	8	9	9	3	3	3
4 ≡ 8	5	3	3	9	5	3	3	9	9	9	3	3	3
3 ≡ 5	3	3	3	9	3	3	3	9	9	9	3	3	3
9 ≡ 3	3	3	3	3	3	3	3	3	3	3	3	3	3

In this example only 18 names are changed during all *unions*, while the unweighted algorithm does 27 name changes.

An alternate approach is the *quick-union* algorithm (again using Yao's terminology) in which the *union* operation is done efficiently, but some *finds* could be expensive. The method is to weaken the requirement that $a[i]$ contain the "name" of the set containing element i by having it contain the index of some other element in the same set, with no cycles allowed. Thus, each set is represented by a tree, which is stored in the a array. For example, the configuration

i	1	2	3	4	5	6	7	8	9	10	11	12	13
$\alpha[i]$	5	11	13	9	5	3	3	8	10	10	12	13	13

is an array representation of the forest

which means that the equivalence classes are $\{1, 5\}$, $\{2, 3, 6, 7, 11, 12, 13\}$, $\{8\}$ and $\{4, 9, 10\}$. Note that an element i is at the root of a tree if and only if $a[i] = i$: these elements are the obvious choices for the representatives of the equivalence classes.

Now *find* is easily implemented by tracing back through the a array until a root element is found:

$j := x;$ **while** $\alpha[j] <> j$ **do** $j := \alpha[j];$

For *union* (x, y) the same method is used to find the roots of the trees containing x and y, and then one is set to point to the other:

$j := x,$ **while** $\alpha[j] < > j$ **do** $j := \alpha[j];$
$i := y;$ **while** $\alpha[i] <> i$ **do** $i := \alpha[i];$
if $i <> j$ **then** $\alpha[j] := i;$

The following table shows the execution of this quick-union algorithm on our example.

	1	2	3	4	5	6	7	8	9	10	11	12	13
	1	2	3	4	5	6	7	8	9	10	11	12	13
$4 \equiv 9$	1	2	3	9	5	6	7	8	9	10	11	12	13
$2 \equiv 11$	1	11	3	9	5	6	7	8	9	10	11	12	13
$7 \equiv 3$	1	11	3	9	5	6	3	8	9	10	11	12	13
$6 \equiv 3$	1	11	3	9	5	3	3	8	9	10	11	12	13
$1 \equiv 5$	5	11	3	9	5	3	3	8	9	10	11	12	13
$11 \equiv 12$	5	11	3	9	5	3	3	8	9	10	12	12	13
$3 \equiv 13$	5	11	13	9	5	3	3	8	9	10	12	12	13
$9 \equiv 10$	5	11	13	9	5	3	3	8	10	10	12	12	13

$2 \equiv 3$	5	11	13	9	5	3	3	8	10	10	12	13	13
$4 \equiv 8$	5	11	13	9	5	3	3	8	10	8	12	13	13
$3 \equiv 5$	5	11	13	9	5	3	3	8	10	8	12	13	5
$9 \equiv 3$	5	11	13	9	5	3	3	5	10	8	12	13	5

This produces the following tree which shows the equivalence class containing all the elements.

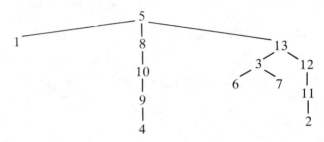

The cost of a *find* operation, the number of array elements examined, is the rank of x in the tree, and the cost of *union(x, y)* is the sum of the ranks of x and y in their trees.

The quick-union algorithm can build unbalanced trees, so there is a corresponding weighted quick-union algorithm that balances the trees by always making the root of the "light" tree point to the root of the "heavy" tree when the union is performed

	1	2	3	4	5	6	7	8	9	10	11	12	13
$4 \equiv 9$	1	2	3	9	5	6	7	8	9	10	11	12	13
$2 \equiv 11$	1	11	3	9	5	6	7	8	9	10	11	12	13
$7 \equiv 3$	1	11	3	9	5	6	3	8	9	10	11	12	13
$6 \equiv 3$	1	11	3	9	5	3	3	8	9	10	11	12	13
$1 \equiv 5$	5	11	3	9	5	3	3	8	9	10	11	12	13
$11 \equiv 12$	5	11	3	9	5	3	3	8	9	10	11	11	13
$3 \equiv 13$	5	11	3	9	5	3	3	8	9	10	11	11	3
$9 \equiv 10$	5	11	3	9	5	3	3	8	9	9	11	11	3
$2 \equiv 3$	5	11	3	9	5	3	3	8	9	9	3	11	3
$4 \equiv 8$	5	11	3	9	5	3	3	9	9	9	3	11	3
$3 \equiv 5$	5	11	3	9	3	3	3	9	9	9	3	11	3
$9 \equiv 3$	5	11	3	9	3	3	3	9	3	9	3	11	3

This is implemented by replacing the last line of the *union* code above by

if $i <> j$ **then**
if $w[j] > w[i]$ **then** $\alpha[i] := j$ **else** $\alpha[j] := i$;

This method clearly produces "flatter" trees than the unweighted method.

Still, none of the above methods are entirely satisfactory: the quick-union and the quick-find methods can be slow, and the weighted methods require extra storage. *Path compression* is a technique which improves the speed without requiring extra storage: after doing a quick-union, make all the nodes on the paths just traversed point to the new root. For example, the equivalence $4 \equiv 8$ transforms

to

Not to

In our example, there is no difference from *quick-union* until the last four equivalences:

	1	2	3	4	5	6	7	8	9	10	11	12	13
		.											
		.											
			.										
$2 \equiv 3$	5	13	13	9	5	3	3	8	10	10	13	13	13
$4 \equiv 8$	5	13	13	8	5	3	3	8	8	8	13	13	13
$3 \equiv 5$	5	13	5	8	5	3	3	8	8	8	13	13	5
$9 \equiv 3$	5	13	5	8	5	3	3	5	5	8	13	13	5

The final tree is better even than that produced by the weighted algorithm, but still not perfecly balanced:

The cost of the *union* operation is multiplied by a constant factor, but the requirement for extra storage is eliminated.

The path compression rule can also be applied to the weighted quick-

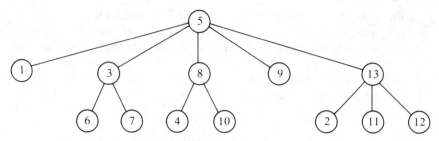

union algorithm, and it could be applied in either algorithm before or after the actual union is done (not to mention during *finds*). These options all lead to methods with different performance characteristics.

The algorithms above are all quite simple and easily coded, but they perform quite differently. It seems clear that the weighted algorithms improve upon their straightforward counterparts, but by how much? And to what extent is path compression an improvement? Unfortunately, not only do we have several possible algorithms to analyze, but also we have several different input models to consider.

12.2 MODELS

To contrast the three models that have been suggested for this problem, we will calculate a precise probability for a small case (the probability, for $N = 4$, that two sets of size two result after two relations have been processed). Also, we will try to see, on an intuitive level, what happens within the models for large N, just before the last few relations which bring together all the elements into one equivalence class are processed. In fact, a goal of the analysis will be to quantify these intuitive statements.

The simplest model is the *random sets* model, used by Doyle and Rivest (1976). Here we consider each pair of classes so far formed to be equally likely to be merged (regardless of size). For $N = 4$, we have three classes after the first relation, one of size two and two of size one. The probability that any two of these will be merged, in particular the singleton classes, is $1/3$. For large N, we have sets of all different sizes, from $O(1)$ to $O(N)$, all equally likely to be merged. This model is generally viewed as the most unrealistic for most applications.

The next model, due to Yao (1976), is called the *random tree* model. Here we consider each way of merging the N items together with $N - 1$ equivalence operations to be equally likely. If we think of the elements as nodes in a graph and the equivalence relations as edges, then a set of edges which merges all the nodes together is a *spanning tree* for the graph. In this model, we first take a spanning tree at random, then use the edges from

that spanning tree in random order. In the fourth tree, the probability that we get two sets of size two after using two edges i 0; in the other three it is 1/3, so the unconditional probability is $(3/4)(1/3) + (1/4)0 = 1/4$. For large N, the most likely situation is that the last few edges will join together large components with $O(N)$ nodes in them.

The third model that we will consider, the *random graph* model, was analyzed by Knuth and Schonhage (1978). Here we suppose that each edge between elements not yet equivalent is equally likely to occur. This is the same as adding random edges to the graph but ignoring those connecting parts that are already connected. For $N = 4$, there are six different edges, with five left after the first is chosen. These occur with equal likelihood, so the probability that the one which connects the two singletons is chosen is 1/5. For large N, the most likely situation is that a large component of size $O(N)$ is quickly formed, and the last few edges just connect single nodes to this component. This model is generally thought to be the most realistic, although, as mentioned above, it is arguable how well actual applications match any of the models.

There are still other possibilities. For example, it is reasonable to do path compression even for calls to *union(x,y)* for which x and y are found to be equivalent. What is needed to model this situation is a *random edge* model, in which each pair $x \equiv y$ is equally likely, independent of the past.

12.3   SUMMARY OF RESULTS

Table 12.1 gives the leading term for the total running time (number of times the a array is accessed) for the various equivalence algorithms in the worst case and in the average case for the three models described above. We will look below at the analyses that produced most of these. With a few very significant exceptions, the constant implied by the O-notation is simply too complicated to fit in the table, but could be worked out. The function $\alpha(N)$ on the last line is an extremely slowly growing function which is constant for practical purposes (see below).

12.4   RANDOM SETS MODEL

The analysis for the quick-find algorithm under the random sets model is the easiest. Let C_N be the total average cost of doing $N - 1$ *union* opera-

TABLE 12.1

	Worst		Average		
			Sets	Tree	Graph
Quick find	$\dfrac{N^2}{2}$		$N \ln N$	$N \sqrt{\pi N/8}$	$\dfrac{N^2}{8}$
Weighted	$N \lg N$		$\frac{1}{2} N \ln N$	$\dfrac{1}{\pi} N \ln N$	$O(N)$
Quick union	$\dfrac{N^2}{2}$		$\left(\dfrac{\pi^2}{3} - 2\right) N$	$\frac{1}{4} N \ln N$	$O(N^2)$
Weighted	$N \lg N$		$O(N)$	$O(N)$	$O(N)$
Compression	$N \lg N$		$O(N)$	?	?
Both	$O(N\alpha(N))$		$O(N)$	$O(N)$	$O(N)$

tions (to form a set of size N). If the last *union* causes sets of size k and $N - k$ to be merged then the total cost is $k + C_k + C_{N-k}$. But the essence of the random sets model is that each value of k is equally likely to occur. This leads to the recurrence

$$C_N = \sum_{1 \le k < N} \frac{1}{N-1} (k + C_k + C_{N-k}),$$

which simplifies to a familiar recurrence (see Chapter 7)

$$C_{N+1} = \frac{N+1}{2} + \frac{2}{N} \sum_{1 \le k \le N} C_k.$$

This is only slightly different from the recurrence arising in the study of Quicksort and binary search trees. The solution is

$$C_N = N(H_N - 1)$$

Analysis of the weighted version involves solving a similar recurrence,

$$C_N = \sum_{1}^{} \le k < N \ \frac{1}{N-1} (min(k, N - k) + C_k + C_{N-k})$$

using precisely the same techniques.

Knuth and Schonage (1978) show how to make the correspondence with binary search trees even more explicit by defining a structure called the *union tree*, which is a full description of a sequence of *union* operations. Defined recursively, the union tree for a sequence of equivalence instructions ending with $x \equiv y$ consists of a left subtree which is the union tree for the equivalence class containing x and a right subtree which is the union tree for the equivalence class containing y. The tree for an equivalence class with only one element is an external node containing that element. The tree for the sequence of instructions used in all the examples above is drawn below.

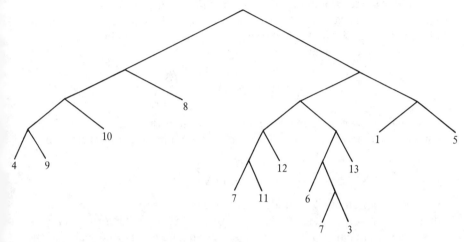

The correspondence with binary search trees is easy to see from this construction. The dynamic process is the same, but run backwards. After the first *union*, we have a forest consisting of $N - 2$ singleton external nodes and one internal node with two external sons. A full tree is eventually built upon these $N - 1$ nodes. If we look at the process backwards we can see that a tree of N nodes comes from replacing an external node by an internal node with two external sons in a tree of $N - 1$ nodes. This is exactly the way that binary search trees are built.

In the union tree, the cost of the last *union* for the quick-find algorithm is the number of nodes in the left subtree (or in the smaller of the two subtrees for the weighted version). For the quick-union algorithm, and its weighted version, the costs are also available as simple properties of the tree, leading to recurrences similar to those above, but which relate the costs of quick-union and quick-find algorithms (see Knuth and Schonhage, 1978). Thus, we will consider only the analyses for quick-find methods under the random spanning tree and random graph models below.

12.5 RANDOM SPANNING TREE MODEL

Much of the machinery above can be applied to the analysis for the random spanning tree model, though the resulting recurrences are much more complicated. The same argument as above shows that the average cost of the quick-find algorithm under this model satisfies the recurrence

$$C_N = \sum_{0 < k < N} p_{Nk}(k + C_k + C_{N-k})$$

where p_{Nk} is the probability that the last *union* operation merges a set of size k with a set of size $N - k$. Knuth and Schonhage give a counting argument (based on the fact that there are N^{N-2} spanning trees on N elements) which proves that

$$p_{Nk} = \frac{1}{2(N-1)} \binom{N}{k} \left(\frac{k}{N}\right)^{k-1} \left(\frac{N-k}{N}\right)^{N-k-1}$$

It is easily shown with Abel's binomial theorem that $\sum_k k p_{Nk} = N/2$ (see Chapter 10), and this leaves, after applying symmetry, the recurrence

$$C_N = \frac{N}{2} + \frac{1}{N-1} \sum_{0 < k < N} \binom{N}{k} \left(\frac{k}{N}\right)^{k-1} \left(\frac{N-k}{N}\right)^{N-k-1} C_k$$

For the weighted quick-find algorithm, we get the same recurrence except that the first term is $\sum_k min(k, N - k)p_{Nk}$. This is similar to the sum for hashing with linear probing that we solved in Chapter 10 and yields to the same techniques: the value is $\sqrt{2N/\pi} + O(1)$.

To finish, then, we need to solve recurrences of the form

$$C_N = X_N + \frac{1}{N-1} \sum_{0 < k < N} \binom{N}{k} \left(\frac{k}{N}\right)^{k-1} \left(\frac{N-k}{N}\right)^{N-k-1} C_k$$

for $X_N = N, \sqrt{N}$. The recurrence becomes somewhat less formidable when both sides are multiplied by $(N - 1)N^{N-1}$ and divided by $N!$

$$\frac{(N-1)N^{N-1}}{N!} C_N = \frac{(N-1)N^{N-1}}{N!} X_N + N \sum_{0 < k < N} \frac{k^{k-1}}{k!} C_k \frac{(N-k)^{N-k-1}}{(N-k)!}$$

From this it can be proven by appealing to Abel's binomial theorem that if

$$C_N = \frac{N-1}{M} \left(\frac{N-2}{N} \cdots \frac{N-M}{N}\right)$$

then

$$X_N = \frac{N-2}{N} \cdots \frac{N-M}{N}$$

for any positive M. A motivation for this solution might be that Abel's theorem provides a way to evaluate sums of the type that appear in the recurrence, so we may as well work backwards, using the theorem to evaluate a general version of the sum for a particular C_N, then solve for X_N. (Knuth and Schonhage give another motivation, based on generating functions.) The result is a family of solutions (parameterized by M) to the recurrence, which leads to a general solution involving linear combinations of members of this family. This solution is expressed in terms of a generalization of the function that we studied in the analysis of hashing with linear probing:

$$Q_i(N) \equiv 1 + \frac{1}{2^{i-1}} \frac{N-1}{N} + \frac{1}{3^{i-1}} \frac{N-1}{N} \frac{N-2}{N}$$

$$+ \frac{1}{4^{i-1}} \frac{N-1}{N} \frac{N-2}{N} \frac{N-3}{N} + \ldots$$

The reader may find it interesting to verify that if

$$X_N = Q_i(N)$$

then

$$C_N = (N-1)Q_{i+1}(N) + NQ_{i+2}(N)$$

This is directly relevant to our analysis for the spanning tree model because $Q_0(N) = N$ (this is a subtle fact which the reader should check). We have proved that

$$C_N = 1/2((N-1)Q_1(N) + NQ_2(N))$$

Now, $Q_1(N)$ is none other than the function that we encountered in Chapter 10:

$$Q_1(N) = \sum_{1 \le i \le N} \frac{N!}{N^i(N-i)!} = \sqrt{\pi N/2} - \frac{1}{3} + O(N^{-1/2})$$

Furthermore, we have the simple bound

$$Q_i(N) < \sum_{1 \le j \le N} \frac{1}{k^{i-1}}$$

which implies, for example, that $Q_2(N) = O(\log N)$ and $Q_3(N) = O(1)$. Putting these results together completes our analysis for the unweighted algorithm:

$$C_N = \sqrt{\pi N^3/8}$$

The analysis for the weighted algorithm proceeds along very similar lines: the solution in this case is

$$\sqrt{2/\pi}((N-1)Q_2(N) + NQ_3(N))$$

The details of working out a more precise asymptotic value for $Q_2(N)$ are left as an interesting exercise for the reader (see Kruskal 1954).

12.6   RANDOM GRAPH MODEL

The full analysis of the performance of the *union* and *find* programs under the random graph model given by Knuth and Schonhage (1978) involves intricate arguments which we do not have room to consider in detail here. It is an asymptotic analysis similar in spirit to those we have been considering, but much more complicated.

As mentioned above, the most likely outcome when edges are added at random to a set of vertices is that one large component is quickly formed, with the last few edges connecting the remaining few single vertices to the large component. This scenario is a classical result proved by Erdos and Renyi (1959). A quick look at the major steps of their proof will not only provide some insights into the behavior of random graphs (and the performance of *union-find* algorithms) but also will illustrate an important method of algorithmic analysis which is applicable to many difficult problems.

There are $\binom{N}{2}$ possible edges connecting N vertices. By a "random graph" we mean one that is formed by adding edges to the vertices in random order (with all $\binom{N}{2}!$ orderings equally likely). How many edges must be added before the graph becomes connected? For the *union-find* algorithms that we have discussed, this point is moot, because we ignored edges within components already connected. However, suppose one wanted to write a program to check the analytic results presented in Table 12.1 by generating random edges, doing a *find* to discover if each should be ignored, then doing a *union* if not. This result will tell us how long such a program will take to complete.

The Erdos–Renyi proof has three parts. First, they show that, after a certain number of edges are added, the probability that there is one very large connected component with at least $N - \log \log(N)$ edges is $1 - O(1/\log(N))$. Next, they show that the log logN remaining vertices are likely to be isolated (with no edges between them) with even higher probability. Finally they calculate the probability that a graph has no isolated vertices. This is asymptotically the same as the probability that a graph is connected, but it is much easier to calculate. (Graphs which are not connected are nearly certain to consist of one large connected component plus some small number of isolated vertices.)

Suppose that C edges have been chosen. The number of graphs with no isolated points can be counted using the principle of inclusion–exclusion. This leads to the following formula for the probability that no isolated points occur:

$$\sum_{0 \le k \le N} (-1)^k \binom{N}{k} \frac{\left[\dbinom{\binom{N-k}{2}}{C}\right]}{\left[\dbinom{\binom{N}{2}}{C}\right]}$$

The crux of the proof is to use Stirling's formula to show that if C is about $\frac{1}{2} N \ln N + cN$ for some constant c then this sum is approximated by

$$\sum_{k \ge 0} (-1)^k \frac{e^{-ck}}{k!} = e^{-e^{-c}}$$

The first two parts of the proof use similar, but more complicated, counting arguments and asymptotic estimates, with the final result that, as $N \to \infty$, the probability that more than $1/2N \ln N + cN$ edges are needed before the graph becomes connected approaches $1 - e^{-e^{-c}}$. This is close to zero for only moderately large c: for example $e^{-e^{-10}} = 0.99995 \ldots$

This proof is an example of the *probabilistic* method of algorithmic analysis: when we can prove that an algorithm is nearly certain to behave in a particular way, then by assuming that it does so we are sometimes led to an easier way to analyze its performance. Of course, the challenge in such a proof is to be sufficiently precise about the likely behavior of the algorithm to be able to make sharp asymptotic estimates: the above proof depends on the discovery of the $1/2N\ln N + cN$ cut-off point. Nevertheless, this is a powerful general technique that must be considered for algorithms which we cannot analyze directly but for whose performance we have some suspicion of a precise description. For example, Guibas and Szemeredi (1978) use this method to analyze the double hashing method that was described briefly in Chapter 10.

From this result we can get some intuitive feeling about the behavior of the *union-find* algorithms under the random graph model. The last few edges connect isolated vertices to a large connected component with about N nodes. For the unweighted algorithm, this will cost $O(N)$ on the average, while for the weighted algorithm the cost will be only $O(1)$ The challenge in the Knuth–Schonhage proof is to show that these performance characteristics are maintained throughout the execution of the algorithms so that the total cost of the unweighted algorithm is $O(N^2)$ while the weighted algorithm is linear. This turns out to be particularly difficult for the unweighted algorithm: a delicate asymptotic argument is used to prove that the cost is $O(N)$, but the constant of proportionality is not known.

12.7 OTHER MODELS

Despite the fact that the various *union-find* algorithms and probabilistic models provide an interesting review of many of the techniques of algorithmic analysis that we have discussed, it must be pointed out that the analysis of the algorithms of greatest practical interest (those using path compression) still remains open. What is worse, it is apparent that none of the models examined above may be appropriate for this problem, because a practical implementation would do path compression in the *find* routine, so that commands requesting that two already connected elements be made equivalent will have an effect. This is an interesting open problem, but it also makes a sobering concluding comment on the importance of the proper choice of model in the analysis of algorithms.

REFERENCES

ABRAMOWITZ, M. and STEGUN, I. A. (1972). "Handbook of Mathematical Functions". Dover, New York.

BENDER, E. A. (1974). Asymptotic methods in enumeration. *SIAM Review*, **16**, (Oct).

DEBRUIJN, N. G. (1958). "Asymptotic Methods in Analysis". North-Holland Publishing Co., Amsterdam.

DOYLE, J. and RIVEST, R. (1976). Linear expected time of a simple union-find algorithm. *Information Processing Letters*, **5**.

ERDOS, P. and RENYI, A. (1959). On random graphs, I. *Publicationes Mathematicae*, **6**.

FLAJOLET, P., RAOULT, J. C. and VUILLEMIN, J. (1977). On the Average Number of Registers Required for Evaluating Arithmetic Expressions. Proc. 18th Ann. Symp. Fnds. Comp. Sci.

FLAJOLET, P. and ODLYZKO, A. (1980). The Average Height of Binary Trees and Other Simple Trees. Proc. 21st Ann. Symp. Fnds. Comp. Sci.

MATHLABGROUP (1977). MACSYMA Reference Manual, Laboratory for Computer Sciences, MIT.

GUIBAS, L. and SEDGEWICK, R. (1978). A Dichromatic Framework for Balanced Trees. 19th Annual Symposium on Foundations of Computer Sciences, A Decade of Progress: 1970–1980. Xerox Palo Alto Research Center, Palo Alto, CA.

GUIBAS, L. J. and SZEMEREDI (1978). The analysis of double hashing. *J. Computer and System Sciences*, **16** (April).

KNOPP, K. (1945). "Theory of Functions". Dover, New York.

KNUTH, D., DEBRUIJN, N. and RICE, S. O. (1969). On the height of planted plane trees. *In* "Graph Theory and Computing", R. C. Reed (ed.).

KNUTH, D. E. (1971) Mathematical Analysis of Algorithms. IFIP Congress. Lubyiana, Yugoslavia.

KNUTH, D. E. (1973a). "The Art of Computer Programming, Volume I: Fundamental Algorithms", 2nd Edition. Addison-Wesley, Reading, MA.

KNUTH, D. E. (1973b) "The Art of Computer Programming, Volume III: Sorting and Searching". Addison-Wesley, Reading, MA.

KNUTH, D. E. (1976). Big Oh and Big Omicron and Big Theta. *SIGACT News*.

KNUTH, D. E. and SCHONHAGE, A. (1978). The expected linearity of a simple equivalence algorithm. *Theoretical Computer Science* **6**, (3), June.

KNUTH, D. E. (1980). "The Art of Computer Programming. Volume II: Seminumerical Algorithms", 2nd edition. Addison-Wesley, Reading, MA.

KRUSKAL, M. D. (1954). The expected number of components under a random mapping function. *Am. Math. Monthly*, **61**.

ODLYZKO, A. (1979). On the number of 2–3 trees. *Theoretical Computer Science*, **10**.

RAMSHAW, L. and FLAJOLET, P. (1980). Grey code and the odd-even merge. *SIAM. J. Computing*, April.

ROBSON, J. (1979). The average height of binary search trees. *Australian J. Computer Science*, **11** November.

SEDGEWICK, R. (1977a). Quicksort with equal keys. *SIAM. J. Computing* **6** June.

SEDGEWICK, R. (1977b). The analysis of quicksort programs. *Acta Informatica*, **7**, pp. 327–355.

SEDGEWICK, R. (1978a). Implementing quicksort programs. CACM, **21**, October.

SEDGEWICK, R. (1978b). Data movement in odd-even merging. *SIAM J. Computing*, **7** (2). August.

SEDGEWICK, R. (1980). "Quicksort". Garland Publishing Co., New York. (Reprint of the author's Ph.D. thesis, Stanford University, 1975).

SEDGEWICK, R. and HONG, Z. (1982a). Notes on Merging Networks. In preparation.

SEDGEWICK, R. (1982b). The Asymptotic Behavior of Heapsort. In preparation.

SEDGEWICK, R. (1982c). "Algorithms". Addison-Wesley, Reading, MA. In preparation.

SHEPP, L. A. and LLOYD, S. P. (1966). Ordered cycle lengths in a random permutation. *Trans. Am. Math. Soc.*, **121**.

TITSCHMARSH, E. C. (1951). "The Theory of the Riemann Zeta-Function". Clarendon Press, Oxford.

WHITTAKER, E. T. and WATSON, G. N. (1927). "A Course of Modern Analysis". Cambridge University Press.

YAO, A. C. (1976). On the Average Behavior of Set Merging Algorithms. 8th Annual Symposium on Theory of Computation.

YAO, A. C. (1980). "Analysis of (h, k, l) Shellsort". *J. Algorithms*, **1** (2). June.

Index